The Field Guide for Powerline Workers

The Field Guide for Powerline Workers

Wayne Van Soelen

DELMAR
CENGAGE Learning™

Australia • Brazil • Japan • Korea • Mexico • Singapore • Spain • United Kingdom • United States

The Field Guide for Powerline Workers
Wayne Van Soelen

Vice President, Technology
Professional Business Unit:
Gregory L. Clayton

Director of Learning Solutions:
Sandy Clark

Managing Editor: Larry Main

Acquisitions Editor: Ed Francis

Marketing Director:
Beth A. Lutz

Channel Manager: Erin Coffin

Marketing Specialist:
Marissa Maiella

Marketing Coordinator:
Jennifer Stall

Development: Sarah Boone

Director of Production:
Patty Stephan

Production Manager:
Andrew Crouth

Content Project Manager:
Benj. Gleeksman

For product information and technology assistance, contact us at
Cengage Learning Customer & Sales Support, 1-800-354-9706

For permission to use material from this text or product,
submit all requests online **www.cengage.com/permissions**
Further permissions questions can be emailed to
permissionrequest@cengage.com

Library of Congress Control Number: 2006023164

ISBN-13: 978-1-4180-1487-2

ISBN-10: 1-4180-1487-7

Delmar
Executive Woods
5 Maxwell Drive
Clifton Park, NY 12065
USA

Cengage Learning is a leading provider of customized learning solutions with office locations around the globe, including Singapore, the United Kingdom, Australia, Mexico, Brazil, and Japan. Locate your local office at **www.cengage.com/global**

Cengage Learning products are represented in Canada by Nelson Education, Ltd.

To learn more about Delmar, visit **www.cengage.com/delmar**

Purchase any of our products at your local college store or at our preferred online store **www.cengagebrain.com**

Notice to the Reader

Printed in the United States of America
2 3 4 5 6 7 16 15 14 13

Preface

This book is intended to give a powerline worker (lineman) in the field access to immediate information on procedures needed to carry out daily duties. Topics of discussion include trouble shooting procedures for transformers, voltage regulators, street lights, ferroresonace, capacitors, high voltage and low voltage. The text discusses correct procedures for stringing, rigging, truck and boom operation, live line work and much more. It has specific safety information for line work including placing grounding, job planning, climbing, precautions, trenching, pole fires, lightning, working alone, and more.

This easy to use, quickly accessible, field reference manual enables the powerline worker to have the most accurate information right at their fingertips. There is a requirement by law for an employer to give a worker the information needed to do their job safely. This book goes a long way towards achieving that requirement.

This field reference manual for powerline workers contains information that is necessary for carrying out daily tasks and for troubleshooting everyday challenges. This manual is intended to support a trained and experienced powerline worker. Having a copy of this book in the truck provides a useful and practical source of information to draw on, while providing assurance to the utility worker that needed information is available and at hand while in the field. It is a companion manual to the *Guidebook for Linemen and Cablemen* which covers the electrical theory training information needed by powerline workers.

While the fundamentals are the same, electrical utility work practices and management systems vary somewhat from place to place. The information is written in terms that promote the reduction of risks instead of quoting rules and regulations. The intention for this field manual is to provide information for the powerline worker so that they are able to work using knowledge that is beyond the requirements of regulatory authorities. This manual can help standardize methods and terms across the country which will aid communication between powerline workers restoring power at neighboring utilities after a storm. This manual is generic enough to be used by all electrical utilities, however, by contacting the publisher, this reference manual can be customized for a specific utility that would like to add their unique procedures and rules into their own version of the manual.

This field manual does not include procedures that can normally be planned in advance or that can be looked up in existing reference material before going to the job (i.e. planned maintenance of a switch or changing a timber on a transmission line using live line tools).

To meet the requirements of ensuring only competent workers perform certain tasks, some utilities conduct annual tests. If this field manual is always readily available to a powerline worker, an open book test would be a legitimate indication of competency. This field manual is the most accurate, up-to-date, and most reliable source of necessary information that a powerline worker needs in order to complete the job in a correct, safe, and timely manner.

Acknowledgements

Delmar, Cengage Learning and the author would also like to thank the following reviewers for their valuable suggestions and expertise:

Alan Drew
Northwest Lineman College

Bob Burton
New York Power Authority

Dan Dade
American Line Builders Joint
Apprenticeship and Training
Program (ALBAT)

James Jones
Oklahoma State University College

Josh Chard
Altel Industries, Inc.

Contents

CHAPTER 3

Operating Switchgear

CHAPTER 4

Trouble Investigation: No Power

CHAPTER 5

Trouble Investigation— Voltage Problems

CHAPTER 6

Transformer Connections and Trouble Investigation

CHAPTER 7
Trouble Investigation: Outdoor Lighting 109

CHAPTER 8
Working with Underground Systems 121

CHAPTER 9
Installing Personal Protective Grounds 135

CHAPTER 10
Stringing Overhead and Pulling Underground 161

APPENDIX 10
Formed Tie Wires. 178

CHAPTER 11
Overhead Connections and Splices 185

APPENDIX 11
Conductor Data 196

CHAPTER 12
Preparing for Hot Work 207

CHAPTER 13
Installing and Removing Revenue Meters. 229

CHAPTER 14
Using the Truck and Boom. 241

CHAPTER 15
Rigging in Powerline Work. 271

CHAPTER 16
Controlling Tree Work Hazards 293

CHAPTER 17
Entering, Inspecting, or Operating in a Substation 317

CHAPTER 1

Essentials for Powerline Workers

1.1 Controlling Hazards in Line and Cable Work

1.1.1 Risk in the Working Environment of a Powerline Worker

The working environment of a powerline worker is very unforgiving. Line work is a high-risk job. Line workers have a high exposure to the potentially fatal hazards of electrical contact, falling from a height, asphyxiation in a vault, or being struck by wayward traffic bursting into a works zone on streets and roadways.

The definition of risk shows why powerline work is considered a high-risk occupation.

$$Risk = Consequences \times Exposure \times Probability$$

Where *Consequences* = the damage or injuries caused when an accident occurs

Exposure = the amount of time a person is within a hazardous area

Probability = the likelihood of making contact with a hazard

Powerline work is considered high-risk work because the *consequence* of a powerline worker accident is often severe. Note that risk is more than merely the chance or odds of having an accident. Not a lot can be done about consequence or *exposure*. Risk is reduced for a powerline worker by using controls (barriers), such as safety equipment that will lower the *probability* of an incident or accident occurring.

High-risk hazards must be identified before starting a job. Any hazards that are identified must be eliminated or controlled. Using a hazard identification list (see Figure 1–1) helps identify hazards that may otherwise be overlooked.

1.1.2 Job Planning and Job Briefings

The most effective means of reducing the risk of an accident is for a powerline worker or crew to have a job plan and to discuss the job plan in a job briefing before starting work.

A good job plan must be written down using a planning format that lays out the job sequence and identifies high-risk hazards and effective barriers for each step, while keeping the amount of writing to a minimum. The job plan shown in Figure 1–2 is an example of one such format.

The foundation of a good job briefing is a discussion, on the job site, structured around a daily job briefing checklist or written job plan. Figure 1–1 is a sample job briefing form that focuses on high-risk hazards and the barriers needed to control the hazards. Similar forms can be created for specialized underground cable work, transmission line work, barehand work, and so on.

Figure 1–2 illustrates a format for planning smaller jobs.

Tailboard Conference Plan

Prepare, discuss, and review the job plan with the crew. Use this form daily and whenever a change is introduced to the job.

Job Being Performed:

Date	Crew Members Present

Hazard Identification List

Gravity	Electricity	Mechanical	Kinetic/Vehicular
Falling from a height Falling objects Falling structures Climbing obstructions	Electrical contact Induction/backfeed Static charge Ground gradients Flash potential Boom contact	Equipment failure Lifting with the boom Maximum working loads on rigging Conductor/guy tensions Vehicle stability	Traffic control Driving conditions Moving loads

Have We Considered?

People	Procedures	Hardware/Equipment	Environment
Person in charge Qualification of personnel Job coordination with other work groups Communication Worker fatigue Pedestrian control General public	Isolation of apparatus Adequate grounding Work protection/Hold off Vehicle grounds Distribution standards Confined space entry Emergency rescue procedures	Work equipment Tools and Personal Protective Equipment (PPE) Vehicles Structures Safe loads for rigging Warning devices Physical barriers	Other utilities Weather conditions Soil conditions Lighting conditions Work schedules

Major Hazards	Barriers to Eliminate or Control

How will we execute a rescue? ____

Exact location for emergency aid: ____

Figure 1–1   Tailboard Conference Plan.

1.2 Safety Responsibilities

1.2.1 Employer and Employee Responsibilities

Labor law dictates that both the employer and the employee have safety responsibilities and rights. The laws may be worded differently in each jurisdiction, but the basics are the same everywhere.

<table>
<tr><td colspan="3">Job: Tree Fallen on Circuit, Cutouts Open on Two Phases, One Conductor Broken</td></tr>
<tr><td>Job Steps</td><td>Major Hazards</td><td>Required Barriers</td></tr>
<tr><td>1. Take a clearance on the circuit and place grounds.</td><td></td><td></td></tr>
<tr><td>2. Remove tree.</td><td>Tree leaning heavily on line, uncertain of how it will drop if cut.</td><td>Rope the tree and pull it off the line with the truck. Keep everyone clear.</td></tr>
<tr><td>3. Repair broken conductor.</td><td>Potential between ends of broken conductor.</td><td>Install grounds on each side of break and use the intact conductors as a jumper.</td></tr>
<tr><td>4. Remove grounds and surrender clearance.</td><td></td><td></td></tr>
</table>

Figure 1–2   Sample format for planning smaller jobs.

Employer Safety Responsibilities

- An employer has a general duty to provide work and a workplace free from recognized hazards and must also provide standards, rules, and regulations that apply to the work being done.

- An employer must inform employees of their rights and duties under the employer's safety and health program and, of course, must not discriminate against employees who exercise their rights under a labor law.

- An employer must ensure that employees have, maintain, and use safe tools and equipment.
- Employers must maintain a log of all employee accidents requiring medical attention.
- An employer must provide an employee's medical and exposure records upon an employee's request.
- An employer is responsible for providing approved personal protective equipment, tools, and hardware that meet regulatory requirements.

Employee Safety Responsibilities and Rights

Labor laws are written to protect employees. An employee should become very familiar with Occupational Safety and Health Administration (OSHA) regulations so that little doubt exists as to employee responsibilities and rights.

- Employees are responsible for performing their jobs safely and for complying with all occupational safety and health standards and rules.
- Employees must comply with the employer's safety and health program by following their supervisors' instructions, participating in required training classes, and reporting any potentially hazardous conditions to their supervisors.
- Employees must wear personal protective equipment specified by the employer.
- Employees are responsible for inspection, care, and use of safety equipment provided by an employer.
- Employees are responsible for reporting equipment defects and for taking defective equipment out of service.
- Employees must report any job-related injuries or illnesses to the employer and must seek treatment promptly.
- Employees have legal rights and are entitled to protection for safety on the job. One such right is that employees are not allowed to be punished or discriminated against by an employer for such acts as complaining to the employer, union, or any government agency about job safety or health hazards.

1.3 Personal Protective Equipment

Personal protective equipment (PPE) is safety apparel that is worn to provide a last-resort protective barrier. PPE does not prevent an incident but wearing it *may* prevent an injury or reduce the extent of an injury. Employers must specify PPE to meet OSHA performance-oriented standards.

Type of PPE	Specifications	Protection
Head protection	Powerline workers wear class E (electrical), type 1 or type 2, hard hats. When new and clean, such hats would have passed a -20,000volt electrical test. Both type 1 and type 2 hard hats meet the same top-impact standards, but type 2 also has protection in the sides, front, and rear. Some employers are now requiring type-2 hard hats.	Line work involves working around overhead structures, hoisting operations, and many other opportunities for severe head injuries. The electrical rating is, of course, not to be depended on under any circumstance; it is only an indication of backup protection that might make a difference if something goes wrong.
Eye protection	Most employers require that eye protection be worn at all times on the job. Prescription safety glasses, or eye protection that is worn over prescription lenses, are available. Filter lenses that have a shade number are needed for protection from light radiation, such as electric arcs.	Always wear eye protection: 1. During switching, installing, and removing protectivegrounds and while working on a hot primary or secondary circuit for protection from the ultraviolet light of an electrical arc or flash and from flying copper or aluminum particles. 2. When striking steel with steel or working with toughened glass insulators. 3. When exposed to strong alkalies or acids during battery boosting, when cleaning rubber hose, and when refurbishing hot-line tools. 4. When firing on wedge connectors. 5. When working with high-pressure hydraulic hoses and other hydraulic tools. 6. When working with power tools, such as chain saws and drills.

(continued)

Type of PPE	Specifications	Protection
Safety footwear	*Mechanical protective footwear* should have a steel toe cap and a puncture-resistant sole and should meet the ASTM F2412 standard (formerly ANSI Z41.1). Look for an ASTM F2412 marking inside the tongue of the boot. "PT" means "protective toe, and "PR" means puncture resistant.	The employer will specify the degree of impact and compression standards. The three levels of impact standards are I-75, I-50, and I-30. The three levels of compression standards are C-75, C-50. and C-30.
	Non-conductive soles should meet the requirements of ANSI Z41 PT9l (M/F) I-75 C-75 EH. "EH" represents the "electrical hazard" designation. EH boots pass a 14,000-volt test when new.	Work boots with nailed soles or stitching that goes all the way through the sole are very conductive and should *not* be used for work around powerlines.
	Dielectric insulation footwear (DI) meeting ASTM F1117 passes a 20-kV electrical test when new. These boots are typically overshoe footwear.	Some employers require dielectric overshoe footwear when working around trucks when the truck boom is in the vicinity of live circuits.
	Electrical conductive footwear (CD)	Electrical conductive footwear is used for two purposes: 1. *For barehand work from a bucket.* The boots help to bond a worker to the metal grid in the bucket. 2. *To prevent the buildup of static on a person's body* and an electric shock each time a grounded object is touched.
	Chain-saw cut-resistant footwear (CS) must meet ASTM F1818.	Regular chain saw users wear cut-resistant boots that are designed to minimize foot injuries caused by accidental contact with a running chain saw.
	Slip-resistant footwear	The most common aids to traction on ice are devices that can be attached to regular footwear.

(continued)

Type of PPE	Specifications	Protection
Flame-retardant clothing	The U.S. Occupational Safety and Health Act prohibits employees exposed to electric arcs or flames from wearing clothing that could increase the extent of an employee's injuries should an arc occur. The employer specifies the type of clothing required for the work being done.	Wear flame-retardant clothing when exposed to electrical arcs and flashes from work such as switching operations, energizing equipment (transformers), accidental short circuits, working on hot secondary lines, and installing meters. Do not wear clothing made from untreated polyester, acetate, nylon, rayon, or blends of any of them.
Hearing protection	The most common hearing protection for powerline work is the disposable earplug. Pairs of earplugs must be inserted properly (see Figure 1–3). **Figure 1–3** How to insert earplugs.	Powerline workers typical require hearing protection when working with or near a portable two-cycle engine, a small four-cycle engine, a chipper, a compressor, a rock drill, a jack hammer, or a helicopter.
Rubber gloves	Rubber gloves (sleeves) are tools for carrying out hot work on distribution circuits, but as personal protective equipment they are also used as backup protection when something goes wrong.	Rubber gloves are used as PPE in the following situations: 1. When opening and closing air-break switches, circuit breakers, or electronic reclosers at an equipment site.

(continued)

Type of PPE	Specifications	Protection
		2. While using hot-line tools, although this varies among employers. 3. When moving or handling energized underground primary cables. 4. When removing sheath from cables and joints and when opening or cutting cables, unless proven deenergized. 5. When working on energized secondaries and services. 6. When handling ropes and conductors while stringing in the vicinity of live circuits.

1.4 Reduce Risk in the Working Environment

1.4.1 Reduce Risk when Climbing and Working Aloft

1.4.1.1 Fall Arrest Systems

Fact	Details
Falling from a heightis a common cause of fatal accidents to powerline workers.	After electrical accidents, the most common cause of fatal accidents to powerline workers has been falling from a height. Falls can be initiated by many factors over which a powerline worker has little or no control, such as wind, ice, knots, rot, loose hardware, leaning structures, and rigging failure. While a bucket is sometimes a safer alternative, climbing is often necessary. A fall protection system is the most effective means of reducing the risk of falling to the ground.
There are four types of fall protection.	**1. A work positioning system** is a personal positioning system that allows a worker to be held in place while keeping hands free for work. A body belt and pole strap constitute a work positioning system. To have fall protection while changing position on a structure, a fall arrest system is used along with a work positioning system.

(continued)

1.4.1.1 Fall Arrest Systems *(continued)*

Fact	Details
	2. A **fall arrest system** is used to prevent a fall to the ground after a fall has been initiated from a structure. A fall arrest system is passive and activated only after a fall occurs. The harness worn in a bucket is a fall arrest. **3.** A **suspension system** is used to suspend a worker who is being transported up or down vertically. It is used to do tree work, work at transmission suspension insulators, and bare-handed work while suspended by a link stick/live line rope, as well as to lower a worker suspended from a winch from a helicopter. To have complete fall protection, a fall arrest system or a secondary backup must be used. For helicopter work, a backup fall protection system relies on the extensive engineering and maintenance involved in the work method. **4.** A **travel restraint system** is a system in which a worker is tied to an anchoring point to prevent reaching an area where free fall could occur. Utility workers on dams or on top of large station transformers use this system.
Without a fall arrest system, there is no backup *physical* barrier to prevent a fall to the ground.	About 60% of falling accidents to powerline workers occur while climbing and 30% while relocating. After a fall is initiated, only a fall arrest system will prevent a fall to the ground; a quick reactive grab or hug of a pole often is not enough. Practical pole climbing and tower climbing fall protection systems are available. After initial resistance to learning a different climbing method, powerline workers can become very comfortable using fall protection equipment when climbing poles and towers.
A full-body harness is the only acceptable attachment of a body in a fall arrest system.	Safety belts that are worn around the waist are acceptable for a travel restraint system or to stop a fall less than 2 ft. (0.6 m) but not as part of a fall arrest system. In a fall, all of the falling force would be concentrated in the abdominal region, if the belt is worn properly, with the D-ring in the middle of the back at the waistline, probably causing internal damage to the body. A full-body harness distributes falling forces throughout the torso, dispersing the forces through the shoulders and down through the thighs.

1.4.1.2 Specific Hazards when Working from a Ladder

Hazards	Barriers
Electrical contact	• All ladders used by utility workers should be nonconductive, preferably fiberglass. • A nonconductive ladder reduces the risk of injury in case of contact with a live conductor and can provide an insulated platform when working on secondary voltage.
Falling from a ladder	• Face the ladder while climbing. • Maintain at least three points of contact—for example, two feet and one hand or, preferably, belt into the ladder with a pole strap. • Avoid overreaching. Keep hips within the ladder uprights. • Never stand above the fourth rung from the top. • Climb the ladder with both hands free. Use a tool belt or pull up material after climbing. • Belt into the ladder with the pole strap around both rails and one wrap around a rung.
Falling with a ladder	• Stabilize the bottom of the ladder. On grass or gravel, use the ladder feet in the spike position. On hard, stable surfaces, use the feet in the flat-foot position. • Tie the top of the ladder, especially if on an aerial cable. Throw a rope over the aerial cable and use it to help set up and tie the ladder into position. • The minimum length of overlap between two sections of an extension ladder should be 3 ft. (1 m) for ladders up to 36 ft. (11 m) high and 4 ft. (1.2 m) for ladders over 36 ft. (11 m) high. • Set the ladder at a proper angle; about 75 degrees. The ladder angle is correct when a person standing at the foot of the ladder can reach the ladder with outstretched arms. Another rule of thumb: Set the horizontal distance of the foot of the ladder out 1/4 of the working length of the ladder. (see Figure 1–4).
Transferring a from the ladder to a roof and vice versa	• Extend the ladder above the roof by at least three rungs to **provide** handhold. • Transfer to the roof from the side of the ladder, not over the top rung.

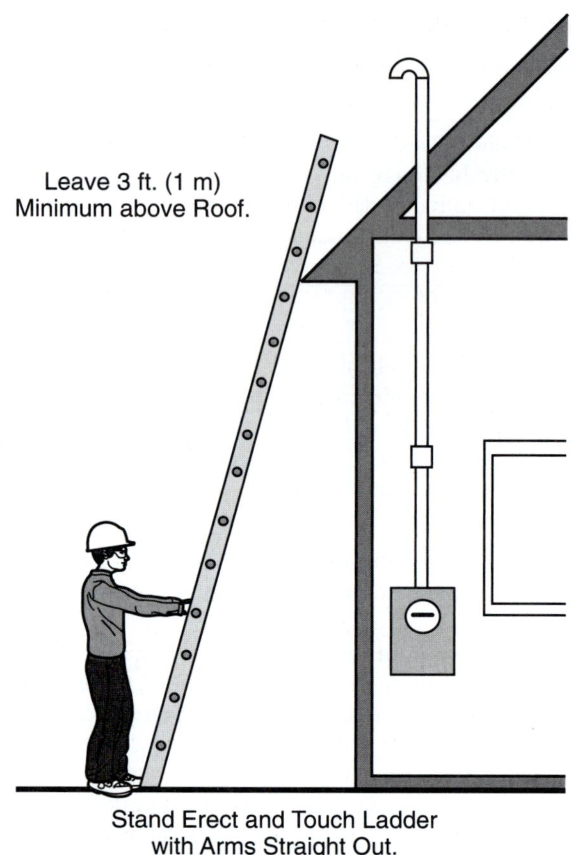

Figure 1–4   Set ladder at proper angle.

1.4.1.3 Reduce the Risk of Falling from a Pole

Hazards	Barriers
A hand line securely fastened to the body belt is snagged by a passing vehicle or by a work vehicle on the job site.	Carry the hand line on a breakaway hook or similar device. Although the probability of a vehicle catching on a hand line may seem rare, it has caused at least one fatality while a lineman was belted into a tower.
A hot conductor is in contact with the pole.	Eliminate possible "pole shock" hazard. Do not climb until the conductor is lifted clear from the pole or the circuit is isolated and grounded.
The pole to be climbed has workers at the bottom of the pole driving ground rods or using a tamper.	Eliminate a possible serious consequence of falling. Do not climb a pole while this work is in progress.
One climber is interfering with another climber.	Wait until the other climber is belted into position before starting to climb.
The pole has knots, weather cracks, rot, or loose hardware.	These climbing hazards can be controlled best by using a fall arrest system.
A powerline worker is climbing while not looking up, increasing the hazard of making contact with a hot conductor.	Confidence in climbing ability allows a line worker to look up while climbing. Use a dedicated observer for novice climbers.
A pole strap can slip over the top of a pole. Exposure to this hazard is increased when climbing a new pole with no hardware on it.	Install at least one bolt near the pole top when setting a pole. This can act as both a physical and a visual barrier. Always belt in below a physical barrier when working near the top of a pole.
The integrity of the hardware supporting a live conductor is suspect and could cause a live phase to fall.	Recognize known hazards such as aluminum-capped dead-end insulators, wood pins, potentially weak conductors such as #4 ACSR or #6 copper, damaged guy wires, and porcelain insulation on switch gear.
Weather-related conditions, such as high wind or ice-coated poles, increase the risk of climbing a pole.	Fall arrest systems are the most effective barriers for these climbing hazards.

1.4.1.4 Reduce the Risk of Falling with a Pole

If	Then
The pole is older than 20 years or is obviously unsafe.	Support the pole with a line truck boom, ropes or guys, or by lashing a new pole alongside it.
The pole species has a history of early failure.	
There are burn marks, woodpecker holes, a large knot, and/or several smaller knots at the same height on the pole.	
The anchor rods and down guys have lost their galvanized coating or are corroded (e.g., near pipeline).	
The pole is beside a trench and/or is in soft, wet, or loose soil.	
There are indications of a shallow setting (e.g., a former ground line above the existing ground level may be seen).	
The strain will be altered while working aloft.	
The pole will be stripped of all conductors and guys. (The pole holds up wires, but the wires also hold up the pole.)	
The pole leans more than 5 degrees.	

1.4.1.5 Working Aloft in a Bucket

When working from a bucket of an aerial device, a fall arrest system must be worn. An approved fall arrest consists of a full-body harness with a lanyard in series with a shock absorber attached to a D-ring on the back of the harness. The other end of the lanyard is attached to a specifically designed anchor point on the boom.

Causes of falls from buckets have included the following:

- Overreaching
- Being ejected from the bucket by a leveling cable failure
- The bucket truck being struck by another vehicle

- The bucket catching on a structure or hardware and suddenly releasing

- Being struck by an object, such as a pole, crossarm, or tree

- Lifting by using the bucket instead of a jib

1.4.1.6 Falling/Dropping Hazards Specific to Tower Work

Hazards	Barriers
Tower climbing hazards include slippery soles, slippery steel, long reaches to the next handhold, frost, ice, wind, or missing a step bolt. A fall from a tower is usually fatal.	Fatal falls have happened to the best climber on a crew. Training and experience are not the best barriers. Fall protection is the only *physical* barrier available to prevent a fall while climbing. Climb up and down towers with a three-extremity contact on the tower at all times. Do not hold onto the step bolts, but slide your hands on either side of the main member at the corner so that you always have hands on the steel.
Objects falling from a tower and striking a worker below can be lethal.	Ensure that no one is under a tower while work is in progress. Workers involved in running a hand line or tag line should work from a position that is clear of any potential falling material.
Material being sent up strikes the tower or spins uncontrolled on the load line.	Tag lines should be used to maintain control of material being raised or positioned. During tower erection, the load line must not be detached from a member or section until the load is secured.
High winds, snowstorms, and ice- or frost-covered steel increase the risk of falling from a tower.	Unless specific training or barriers have been put in place for the hazard encountered, work aloft should not be started during these conditions.
Tripping or loss of balance occurs while carrying tools or material across a tower.	Fall protection is the only *physical* barrier against falling while walking across a tower bridge.
The jolt from a static electric shock while climbing from an insulated ladder onto steel or when contacting a ground wire while climbing a wood structure can result in inadvertent movements.	Being prepared and making a fast slap with the hand to bond onto the grounded object will reduce the shock and the risk of an inadvertent movement. Fall protection is the only *physical* barrier to prevent inadvertent movement.

(continued)

1.4.1.6 Falling/Dropping Hazards Specific to Tower Work *(continued)*

Hazards	Barriers
An insulator string is an unstable platform for climbing, as well as a potential source for a serious gash from a broken insulator.	Use a ladder, platform, or scaffold to work at or beyond an insulator string. Fall protection is the only *physical* barrier available to prevent a fall from an unstable platform.

1.4.2 Working Alone

A general principle for powerline work is that rescue capability must be available when working on circuits over 750 volts, when working at heights of more than 10 feet (3 m), or when working in a confined space.

When a second person is needed, the second person must be trained and equipped to perform applicable rescues: pole-top rescue, emergency bucket rescue, bucket lowering, tower rescue, tree rescue, and confined space rescue. First aid, artificial resuscitation, and/or cardiopulmonary resuscitation (CPR) are some of the qualifications required of a second person.

Typical one-person tasks in some utilities:

- Operating switches by means of an operating handle or switch sticks
- Re-fusing circuits or equipment with a hot stick or extension stick
- Taking voltage and current readings

1.4.3 Utility-Specific Traffic Hazards and Barriers

Hazards	Barriers
Drivers are looking for physical objects and do not see conductors or ropes on the road. Contact with a conductor or rope being strung across a roadway during a *slack stringing operation* is a utility-specific hazard.	Stop traffic *before* running wire or rope across a roadway. Waiting for a break in traffic and then stopping vehicles entering the work zone increases the risk. The driver often does not see any obstruction when there is only a wire suspended across the road.

(continued)

Hazards	Barriers
Drivers are looking for physical objects and do not see conductors or ropes being strung. Contact with a low-hanging conductor or rope being strung across a roadway during *a tension stringing operation* is a utility-specific hazard.	**Caution:** If the road crossing is a much longer span than others in a pull, it is more likely that extra conductor will sag into that span. Rider poles or truck booms can be set up to reduce the risk of conductor coming down into the traffic. Traffic controls and flaggers must stand by to respond when conductor starts to sag too much.
The elbow of an aerial bucket stretches over an open lane of traffic and a contact is avoided only if an approaching truck is alert to the hazard.	Block off enough lanes to cover the potential of the boom elbow over the roadway and/or use a dedicated observer (see Figure 1–5).

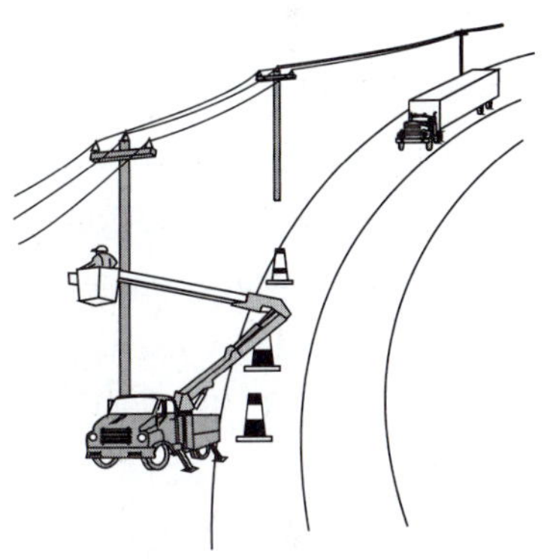

Figure 1–5   Traffic hazard alert.

If traffic is allowed to run over a wire that is being strung across a road, the wire can be picked up by the tires of a fast-moving vehicle.	Stop traffic first, then direct vehicles to drive slowly over wires on a roadway. On high-speed highways, use the police or highway patrol to bring traffic to a rolling stop.
A specific work zone location is very vulnerable to fast-moving traffic.	In addition to cones, flaggers, etc., a physical barrier such as a large truck (crash truck) parked in a closed traffic lane should be considered for use in very vulnerable locations.

1.4.4   Working in a Confined/Enclosed Space

Because most confined spaces for electrical utilities are designed to be entered under normal operating conditions, the Occupational Safety and Health regulations in the United States refer to them as enclosed spaces, rather than Permit Required Confined Spaces. An enclosed space still must be tested for oxygen deficiency, flammable gases, and vapors.

- For workers who enter an enclosed space infrequently, preparing a permit (Figure 1–6) before entering will reduce the risk of overlooking an important step in the procedure.

- Testing is done for oxygen deficiency, flammable gases, and vapors before removing a cover so that a spark created by the opening of a cover or hatch will not cause an explosion. Remote probes are used to check for explosive gases. **Caution:** Flammable gases or vapors may not register during a test until oxygen is introduced into the space.

Confined/Enclosed Space Hazards

Reduce your risks when working in vaults:

- Do not climb into or out of a vault by stepping on cables or hangers.

- Before lowering tools or material into a vault, ensure that everyone below is clear of the area directly under the opening.

- If duct rods are used, someone must be stationed at the far end of the duct line to ensure that the required minimum approach distances are maintained.

- When multiple cables are present, the cable to be worked on must be identified by testing or spiking.

- An impending fault is indicated when a cable or joint has signs of oil or compound leaking, broken sheaths, hot localized surface temperatures, or joints that are swollen. The defective cable should be deenergized before anyone works in the vault.

- When work is performed on cables in vaults, continuity of the metallic sheath must be maintained. A lethal voltage may be present between the ends of a break in a sheath.

Electrical Utility Enclosed Space Entry Permit

Site Location and Description: _________________________________

Supervisor: _________________________________

Date and Time Issued: _________________________________

Date and Time of Expiration: _________________________________

Authorized Entrants: ______________   ______________

______________   ______________

Attendants: ______________   ______________

Identify Potential Hazards	**Control Measures**	**Additional Control Measures**
Inadequate Electrical Clearances	Lockout/Tagout/Ground	_________________
Limited Exits	Set Up Rescue Capability	_________________
Traffic and Pedestrian Control	Set Up Barriers	_________________
Lack of Natural Ventilation	Ventilate	_________________
Airborne Combustible Dust	Ventilate	_________________
CO from Vehicle Exhaust	Ventilate	_________________
CO_2 from Rotting Vegetation	Ventilate	_________________
Nitrogen from Pressurized Cable	Ventilate	_________________
H_2S	Ventilate	_________________
The vault is over 15 ft. (4.5 m) deep	Ventilate	_________________
Hot Work (i.e., soldering, pouring compound, cutting, or heating)	Ventilate	_________________
Other: _______________	Purge—Flush and Vent	_________________

Test and Monitor Atmosphere (record monitoring every 2 hours)

Test to Be Taken	**Limit**	**Completed**	**Time**	**Time**	**Time**
% oxygen	19.5–23.5%	_______	_______	_______	_______
% LEL (LFL)	10% Max.	_______	_______	_______	_______
Carbon monoxide	35 ppm	_______	_______	_______	_______
Hydrogen sulfide	10–15 ppm	_______	_______	_______	_______
Other toxics	PEL	_______	_______	_______	_______

Rescue Readiness

Communications Setup	_____	Lifelines, Hoisting Equipment	_____
Full-body Harness	_____	Explosion-proof Lighting	_____
Protective Clothing	_____	Respiratory Protection	_____
Fire Extinguishers	_____	Special Tools	_____

Other Information/Specific Requirements _______________________________

This information and the work covered by this permit have been reviewed with all entrants and attendants. Safety procedures have been received and are understood. All appropriate items have been completed.

Entry Authorized _______________________________

Title _______________________________

Date/Time _______________ / _______________

THIS PERMIT SHOULD BE KEPT POSTED AT THE JOB SITE.

Figure 1–6 Typical electrical utility enclosed space entry permit.

1.4.5 Working in Excavations and Trenches

Hazard	Barrier
Digging into underground utilities can lead to explosions, flooding, or other high-cost damage.	Obtain locations for gas, water, sewer, telephone, TV cable, and powerlines. When in doubt, dig by hand or use a heavy-duty industrial vacuum loader (sucker) truck that pneumatically sucks up solids, sludges, and slurries. Check for utility markers. *Utility markers* Red — Electric Yellow — Gas, Oil, Steam Orange — Communications, Cable Blue — Water Green — Sewer Purple — Irrigation White — Proposed Excavation Pink — Temporary Survey Tape
A deficiency of oxygen or the presence of other hazardous gases can be dangerous in an excavation.	Carry out air sampling in trenches more than 4 ft. (1.2 m) deep where oxygen deficiency or other hazardous atmospheres could reasonably be expected to exist.
A cave-in can bury a worker.	• Reduce the risk of cave-ins from material that could fall or roll from an excavation face or from material piled next to an excavation. • Ensure a proper slope angle for the type of soil being excavated (Figure 1–7). Refer to

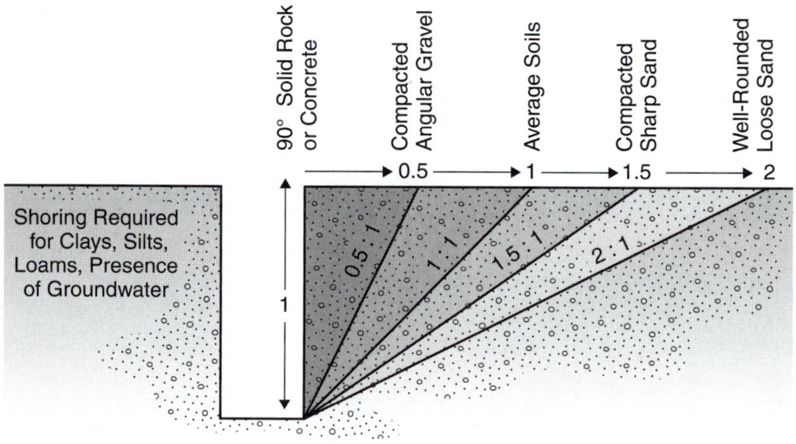

Figure 1–7   Slope angles of a trench.

(continued)

1.4.5 Working in Excavations and Trenches *(continued)*

Hazard	Barrier
	applicable regulations regarding the minimum slope, bracing, and piling needed for the various types of soil in your work area.
	• Place a protective system, such as a trench box (or trench shield), when exposed to the hazard of falling or sliding material from an excavated bank or side more than 4 ft. (1.2 m) high above a worker's feet (Figure 1–8).

Figure 1–8 Protective barriers for trenching.

Hazard	Barrier
	• Keep excavated material and work equipment at least 3 ft. (1 m) from the trench to prevent overloading and stress cracks.
	• Leave a trench during a sudden downpour because the rain can fill a trench and cause rain-soaked soil to give way.
A worker is on the receiving end of a stone "shot" from between a truck tire and pavement.	Have loose stones swept from any work area.
During a cave-in emergency, workers are uncertain about what to do.	Have an emergency plan. Always provide a stairway, ladder, or ramp for a hurried escape route at a distance of no more than 25 ft. (7.5 m) in any direction. When caught in a cave-in, run or jump *up* the bank, not down.

1.4.6 Working in Heat and Cold

On the hottest or coldest days of the year, transformers tend to burn out from overload.

A human body is most efficient within a temperature range of $+/-3°F$ ($+/-2°C$) of the body's normal temperature of 98.6°F (37°C). The internal human thermostat works very well but will be stressed or overwhelmed when working outdoors on very hot days or very cold nights.

1.4.6.1 Working in the Heat

Symptoms of heat stress are headaches, dizziness, lightheadedness, irritability, confusion, upset stomach, fainting, and/or pale, clammy skin. Heat stress left untreated could lead to a fatal heat stroke. Flame-retardant clothing may increase the risk of heat stress; therefore, the following precautions become more critical.

To avoid heat stress, do the following:

- Before you get thirsty, drink a lot of water.

- Wear light, loose-fitting, breathable clothing, such as cotton.

- In extreme heat, take short breaks in air-conditioned vehicles or buildings or in the shade.

- Eat smaller meals and avoid caffeine and sugar.

1.4.6.2 Working in the Cold

Symptoms of cold stress that could lead to dangerous hypothermia are severe shivering, slurred speech, clumsiness, poor judgment, confusion, apathy, slow pulse or slow breathing, excessive fatigue, drowsiness, reduced sense of touch, less grip strength, and less ability to sense heat, cold, and pain.

To avoid cold stress, do the following:

- Have regular warm-up breaks in a heated truck cab or indoors with an opportunity to remove clothing to prevent sweating. A person who has become damp or sweaty will chill quickly and be susceptible to hypothermia.

- Hot drinks will provide energy and warmth and will prevent dehydration.

- Use a buddy to check each other's faces for frostbite.

- Dress for the cold by wearing layers, including an outer shell to protect from the wind and inner insulated layers. Layers can be added and removed to stay comfortable and to prevent dampness.

- Wear a good winter hard-hat liner covered with a parka hood; half a person's heat loss can be through the head.

1.4.7   Lightning in the Vicinity of Work

A voltage surge from lightning travels like an ocean wave in all directions, except that it travels at the speed of light. The wave is diminished when it encounters a location where it will flash over, preferably at a surge arrester, where some or all of the energy is dissipated.

A voltage surge causes a coupling effect between the phase conductor and the neutral or shield wire. The coupling effect reduces the voltage on the phase conductor while it raises the voltage on the neutral or shield wire.

There can be a rise in voltage on a neutral or shield wire at some distance from a thunderstorm. When lightning is seen in the distance, workers should get clear of electrical circuits.

The following safety tips can apply to on-the-job and off-the-job situations:

- The safest place to be during a thunderstorm is in a vehicle. When a vehicle is struck by lightning, all the metal in the vehicle is at the same potential, so anyone in the vehicle would not be exposed to any potential difference.

- When your hair stands on end and your skin starts to tingle, a potential difference is building up between the ground you are standing on and the clouds above. Lightning is about to strike. Drop into a fetal position.

- Usually lightning strikes tall objects such as trees. People nearby are injured due to the ground (voltage) gradients of the lightning dissipating into the ground. Each gradient is at a different voltage, with the greatest potential differences being near the center. Staying away from tall objects reduces exposure to the highest voltage gradients. Keeping your feet together reduces exposure to the different potentials between gradients.

- Grounded objects will rise in potential when a lightning strike is nearby. Inside buildings, stay away from grounded objects.

1.4.8 Stay Alert for Public Electrical Contact Hazards

Powerline workers are the most qualified people to identify potential public electrical contact hazards with utility circuits and should report any of the following:

- Flagpoles, antennas, and ladders within striking distance of an electrical circuit

- Cranes, boom trucks, and ladders within 10 feet (3 m) of a hot distribution circuit

- Construction and buildings being erected within established safe limits

- Trees near powerlines, such as those in suburban streets, school yards, and parks, that are liable to be climbed by young people

- Farm equipment, such as grain augers and high loads, being moved under lines

- Irrigation systems with large solid streams of water near transmission lines

- House movers traveling under powerlines

- Digging activity or posts or bars being driven into the ground near transmission or distribution cables

- Irrigation pipes stored underneath or near powerlines

1.5 Managing Emergencies

1.5.1 Managing and Communicating Emergencies

A utility/employer must have in place plans, procedures, and training for such emergencies as an electrical system failure, fire, injuries on the job, and rescues from aloft or from a confined space.

Lessons learned that have evolved into specific rules include the following:

- A job briefing on emergency measures should be provided, including written documentation of emergency telephone numbers and the type of rescue that personnel should be ready to perform.

- A hand line should be hung anytime a person is working aloft. The rope and hardware must have a 10-to-1 working load factor and undergo scheduled inspections.

- A live line cutter should be readily accessible on a pole during any hot work.

- Emergency code words should be designated that will open up a radio and give a control center the authority to isolate a circuit without question.

- Rescue packs should be available on every aerial lift, regardless of whether or not the baskets tilt (it may not always be possible to lower the baskets all the way to the ground).

- Keeping the deck of a bucket truck free of hardware, especially near the lower hydraulic controls, should be a priority.

1.5.2 First Aid Summary

The purpose of first aid is to preserve life and apply measures to stabilize the effects of an injury. Take first aid and CPR training. Be sure that you know the emergency radio procedures for your utility.

In summary, look for and apply aid for the three B's in the following order of priority:

1. Breathing

2. Bleeding

3. Bones

1.5.3 CPR Summary

An electric shock can put the heart into fibrillation. When that happens, individual heart muscle fibers do not work together and the blood no longer circulates efficiently. Cardiopulmonary resuscitation (CPR) can be used to circulate oxygenated blood to the brain manually until medical aid is available. The following CPR summary calls for two breaths for every 15 chest compressions. Now some experts suggest 30 compressions between breaths. Use the methods for which you have received training.

1. Check for response.
   - Does the casualty respond to voice or painful stimulus?
   - If *yes,* check the casualty for other conditions and call for help if necessary.
   - If *no,* shout for help and continue.
2. Check for breathing.
   - Open the airway, tilt back the head, and lift the chin. Is the casualty breathing?
   - If *yes,* place the casualty in the "recovery position" and call for help.
   - If *no:*
3. Give two effective rescue breaths.
   - Are you alone?
   - If *no,* ask a helper to call an ambulance.
   - If *yes,* carry out resuscitation for 1 minute, then call for an ambulance.
4. Look for signs of circulation.
   - Are there signs of circulation?
   - If *yes,* continue rescue breaths: 10 breaths a minute for an adult or 20 breaths a minute for a child. Repeat step 2 after every set of breaths.
   - If *no:*
5. Start CPR.
   - For an adult, alternate 15 chest compressions with 2 breaths.
   - For a child or infant, give 5 compressions to 1 breath.
   - Continue CPR until emergency help takes over, the casualty moves or takes a breath, or you are too exhausted to continue. CPR is intended to artificially pump oxygenated blood to the brain until a heart defibrillator can be used. Some utilities/contractors keep an automated external defibrillator (AED) on designated trucks to be used by trained personnel (see Figure 1-9).

Some utilities/contractors keep an automated external defibrillator (AED) on designated trucks to be used by trained personnel.

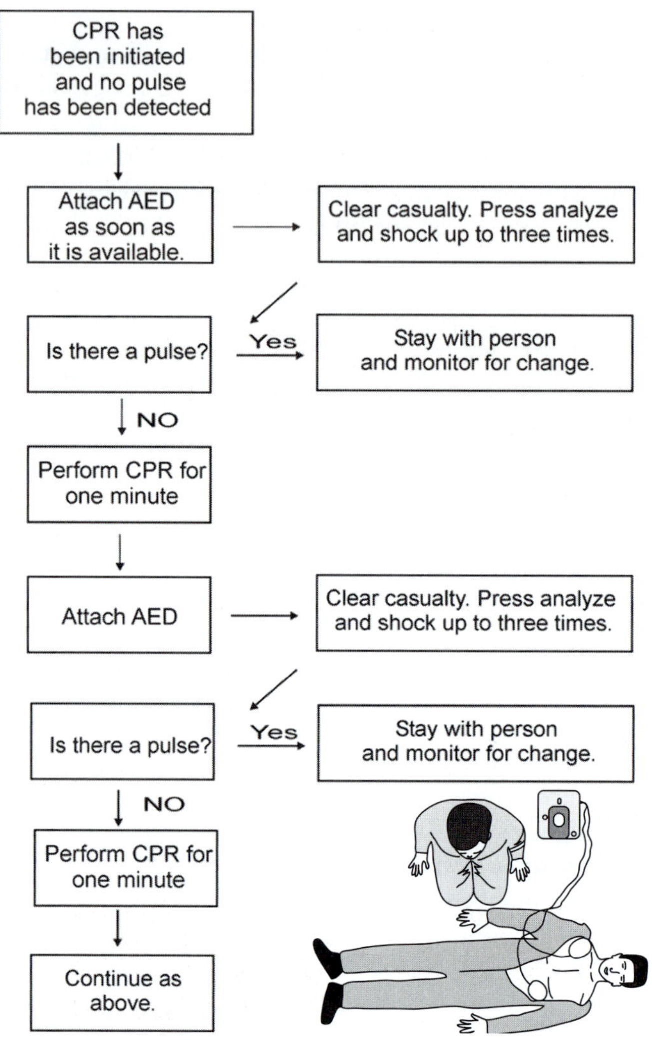

Figure 1–9   Automated external defibrillator.

1.5.4 Seek Medical Attention for Electric Shock

The extent of damage from electric shock is not always visible. After initial first aid treatment, it is prudent to seek medical attention if any of the following is observed:

1. There is or was any loss of consciousness.

2. There is any burn mark. A burn mark indicates that electric current has entered and exited the body and has done unknown damage to internal tissue and organs.

3. The victim has an irregular heartbeat.

4. The victim experiences persistent pain and/or anxiety about the nature of the injury.

1.5.5 Fighting a Fire in an Electrical Environment

Isolating a circuit in or near a fire creates the safest environment for line crews and firefighters. Check for hot conductors lying on the ground. Water flowing near hot conductors will be conductive and hazardous to anyone walking on wet ground.

A *transformer fire* is actually a flammable liquid fire with the added hazard of it being in an electrical environment. However, if a transformer is on fire, the transformer is not salvageable. Stay upwind—if the transformer oil is contaminated with polychlorinated biphenyls (PCBs), the smoke has some very dangerous chemicals in it.

A *pole fire* is a wood fire in an electrical environment. Be aware, however, that any spent dry powder lying on a crossarm or on other hardware can become contaminated and conductive, especially in damp conditions. Wet powder could become a path to ground and cause an explosive arc near any worker trying to put out a fire.

Figure 1–10 shows the three ratings important to powerline workers when fighting a fire around powerlines. Any extinguisher used in an electrical environment must have a C rating in combination with another rating. The extinguishers normally carried on electrical utility trucks are rated as ABC, which can be used for transformer fires, pole fires, or vehicle fires.

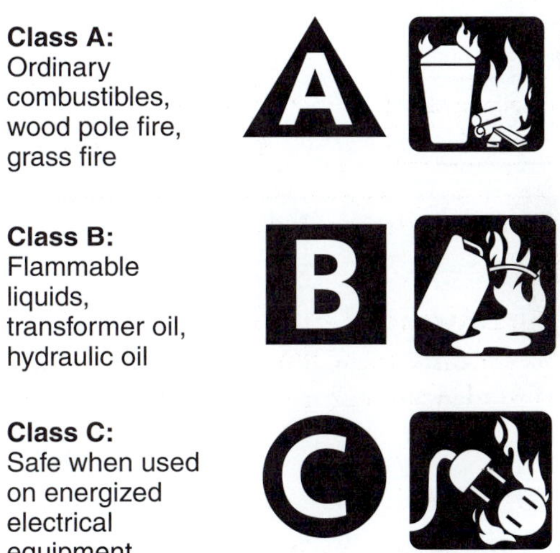

Figure 1–10   Types of fire extinguishers.

1.5.6 Electrical System Emergencies

When storms hit and an electrical system goes down, powerline workers go to work.

Some notes for powerline workers:

- A major storm with many utility customers out of power is a crisis but not an emergency. Any institution or person that has a life-and-death need for uninterruptible power has the responsibility to have backup generation. Line work must still be planned, job briefings carried out, and complete lockout/tagout procedures used for an isolation guarantee.

- Many utilities insist that any outside crews work to the host utility standards. Some utilities provide an orientation and a booklet for outside crews with a brief description of the voltages, the type of system control, the hours of work, fusing requirements, and much more.

- Ideally, the host utility will give an outside crew work assignments that are confined to a circuit or specific geographic area where the work is independent of other crews. Line work requires everyone on a job site to know the same job plan, take part in the same job briefing, use procedures that will not conflict with others, and stay within the confines of one isolation guarantee covered by a lockout or tagout.

- Fatigue from lack of sleep will be an issue. Rest during the delays that frequently occur in the administration of outside crews. Ideally, work will be scheduled to take advantage of all the daylight hours available.

Electrical Awareness for Line Work

2.1 Typical Effects of Electrical Current on the Human Body

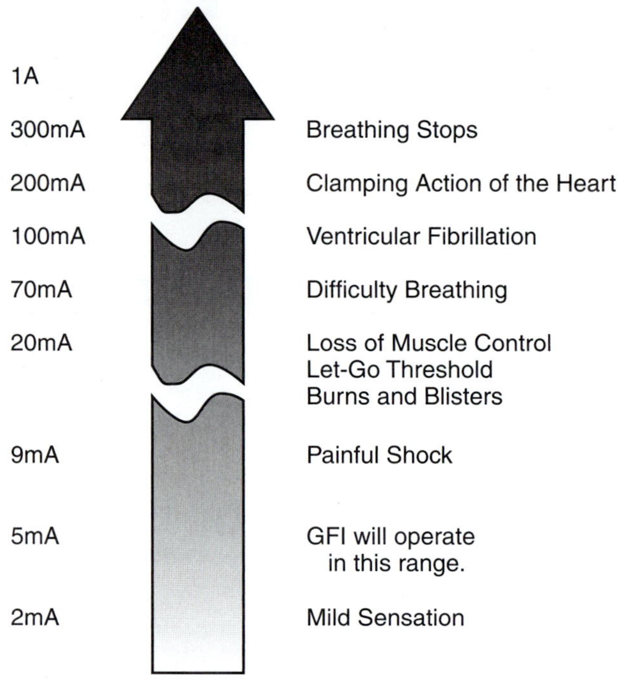

Figure 2–1 Typical effects of electrical current on the human body.

2.2 Applying the Minimum Approach Distance

Maintain a safe distance to exposed live apparatus. In most jurisdictions, occupational safety regulations do not allow an unqualified person to come within 10 feet (3 m) of a distribution electrical circuit and an additional 4 inches (10 cm) for every 10 kV over that, which works out to 14 feet (4.3 m) for 169 kV, 16 feet (4.9 m) for 230 kV, and 25 feet (7.6 m) for 500 kV.

The minimum approach distances in Table 2–1 (which is OSHA Table S-5) apply to qualified people working in the vicinity of bare live apparatus.

2.3 Electrical Contact Hazards

2.3.1 Series Contact

When a power conductor is cut, all the current flow is interrupted. The resistance across the open point is infinite; therefore, the voltage drop (IR) is at full-line voltage. This full-line voltage (recovery voltage) appears across the two ends of the cut conductor. Install a jumper across an open point every time a *cut* or *joint* is made to a current-carrying conductor. The current keeps flowing, but the voltage across the cut is at 0 volts.

TABLE 2–1   **Approach Distances for Qualified Employees**

Voltage range (phase to phase)	Minimum approach distance
300V and less	Avoid contact.
Over 300V, not over 750V	1 ft., 0 in. (30.5 cm)
Over 750V, not over 2kV	1 ft., 6 in. (46 cm)
Over 2kV, not over 15kV	2 ft., 0 in. (61 cm)
Over 15kV, not over 37kV	3 ft., 0 in. (91 cm)
Over 37kV, not over 87.5kV	3 ft., 6 in. (107 cm)
Over 87.5kV, not over 121kV	4 ft., 0 in. (122 cm)
Over 121kV, not over 140kV	4 ft., 6 in. (137 cm)

Source: OSHA Table S-5.

2.3.2 Parallel Contact

In a *wye circuit,* all objects—including, live-line tools, rubber gloves, and insulated booms in contact with both the ground and a live conductor—form a parallel connection on the circuit and will be subject to some leakage current.

When a parallel contact is made, keep the current flow through the body-to-ground below a dangerous threshold by using rubber gloves, live-line tools, insulated platforms, and ground-gradient mats where applicable.

2.4 Electrical and Magnetic Induction Hazards

2.4.1 Electric Field (Capacitive) Induction Hazards

Hazard	Barrier
Contact with an isolated circuit subject to electrical field (capacitive) induction can result in a lethal steady-state shock that continues as long as contact is maintained.	Installing a set of grounds on the conductor discharges the induced voltage effectively. Bonds between vehicles, ground probes, conductors, and so on eliminate exposure to potential differences among them.
When climbing a transmission tower or working in a high-voltage substation, a person's body acts as the second plate	Wearing conducting-sole boots reduces this hazard when working on transmission towers or working in high-voltage stations.

(continued)

2.4.1 Electric Field (Capacitive) Induction Hazards *(continued)*

Hazard	Barrier
of a capacitor. When a grounded object is touched, the resultant spark consists of a high-current discharge that could result in an involuntary movement or fall.	
A person using rubber gloves on a high-voltage distribution line, such as a 34.5kV line, might feel a vibration or "bite" in those gloves.	The rubber glove acts as insulation between two conductive plates, the first being the 34.5kV and the second the worker. This effect is not considered a major hazard.

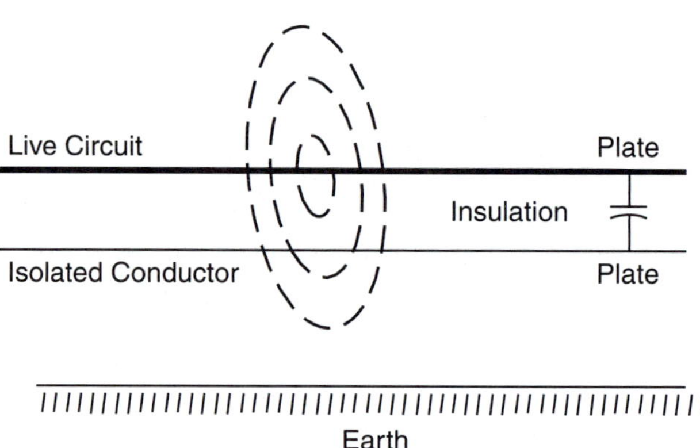

Figure 2–2   Capacitive effect between conductors.

2.4.2 Magnetic-Field Induction Hazards

Hazard	Barrier
An isolated and grounded conductor paralleling a live current-carrying conductor has current induced on it. As illustrated in Figure 2–3, the installation of two sets of grounds (bracket grounds) on the isolated conductor creates	When only one set of grounds is installed at the point of work, a circuit in which an induced current would flow has not been created. When the conductor is grounded in two locations, a circuit is created and current flows through the conductor and through the grounds. If this circuit is interrupted by removing a ground or cutting a conductor, a high voltage is available across the

(continued)

2.4.2 Magnetic-Field Induction Hazards *(continued)*

Hazard	Barrier
a circuit with current flowing along the conductor, down one set of grounds through earth (or neutral), and up the other set of grounds. Because the amount of induction from a magnetic field depends on current and not voltage, this induction is also a hazard on lower-distribution voltages.	open point. That is why it is important to install and remove grounds with a live-line tool.

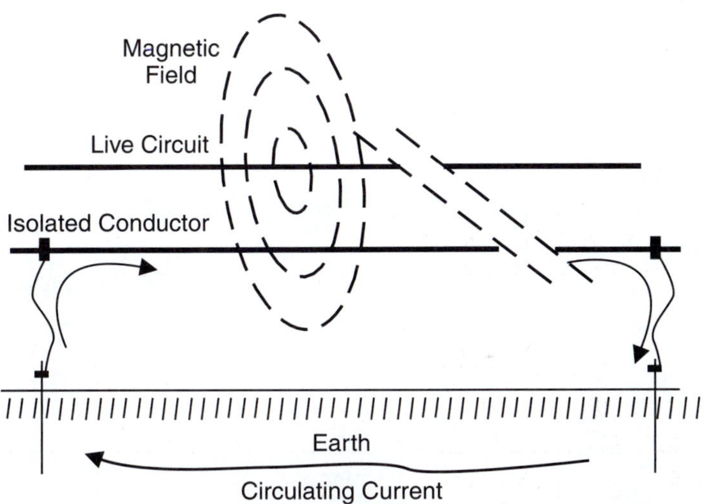

Figure 2–3   Magnetic field effect between conductors.

2.5 Voltage Gradient Hazards

Hazard	Barrier
Anyone working at a location where a ground fault enters the earth is subject to ground-gradient voltage or step potentials.	One form of protection from ground gradients is to work from a ground-gradient mat (an equipotential mat) that is bonded to the device (grounded transformer tank, switch handle, cable sheath and so on) that may become energized.

(continued)

2.5 Voltage Gradient Hazards *(continued)*

Hazard	Barrier
Ground potentials can occur when setting a pole in a live circuit, switching, working at a pad mount or submersible transformer, working near a ground rod, and so on.	Figure 2–4 shows a person standing on a ground-gradient mat that is electrically bonded to the switch being operated. If the cap and pin break off the post insulator and energize the steel structure, the person will be safe because no voltage difference would exist between the hands and the feet. Figure 2-5 shows a person working on a ground-gradient mat that is bonded to a pad mount transformer. If a ground fault on a cable or elsewhere raises the potential on the transformer tank, the person would be safe because no potential difference would occur between the hands and the knees.

Figure 2–4 Operating a switch from a ground mat. *Source:* Kri-Tech Power Products Ltd.

Figure 2–5   Working at a pad-mount transformer from a ground mat. *Source: Kri-Tech Power Products Ltd.*

2.6 Hazards When Working with a Neutral

Hazard	Barrier
Although no apparent voltage exists on a neutral, a high voltage can occur across a break or open point neutral. Current flow is interrupted, creating an infinite resistance, and all the voltage (recovery voltage) pushing the current appears across the open point. This is particularly hazardous during a coincident system fault.	It is a basic lines procedure to always jumper a neutral or cable sheath (or any other current-carrying conductor) before it is cut or joined.
A very high voltage can occur on a neutral when a phase is struck by lightning. The high voltage on the phase couples with other phases and the neutral. Electrical coupling	Do not work on a neutral when lightning is in the vicinity.

(continued)

2.6 Hazards When Working with a Neutral *(continued)*

Hazard	Barrier
with the neutral tends to bring the phase voltage down, but electrical coupling also raises the voltage on the neutral before the voltage is dissipated to ground on the multi-grounded neutral.	
On some types of construction, especially when performing storm work in unfamiliar territory, it is very easy to misidentify the neutral. • On some two-phase delta construction, one phase is in a position that would normally be the neutral position in a wye system. • The neutral can be a primary position. • The neutral can be anywhere on a secondary rack.	Trace the neutral from the closest ground connection or transformer. Placing a ground on the neutral is a positive identification method.

2.7 Establishing an Isolation Guarantee Using a Lockout/Tagout Procedure

Do not approach a hot circuit closer than the minimum approach distance unless you are insulated, the circuit is insulated, *or* the circuit **has been isolated and protective grounds have been installed.** The circuit is isolated and an *isolation guarantee* (commonly called a "clearance" in many places) is obtained based on a formal lockout/tagout procedure. An isolation guarantee is a document prepared by a responsible person that states all of the following:

1. A particular line or piece of electrical equipment has been disconnected from all known sources of energy.

2. All feed points have been opened, tagged, and/or locked.

3. The isolation will remain in effect until formally surrendered.

A lockout/tagout procedure for lines would not necessarily meet the regulatory locking requirements in a generating station. In lines, one person may be signed on as the holder of the isolation guarantee and be responsible to everyone on the job and that for ensuring that the line is isolated and grounded will remain isolated and grounded.

Common Elements	Details
1. A controlling authority (system control or dispatch) is responsible for granting an isolation guarantee on a circuit or equipment.	1. The controlling authority is normally the system control (dispatch) for the system. 2. In some utilities, powerline workers are given control of the distribution system, especially for a switchgear that is not remotely operated. They will act as their own dispatchers and will prepare their own switching orders and will apply tags on applicable switchgears. There are some switchgears to open or close safely, such as a tie point between two substations, that may require information only available to system control personnel.
2. An application is made out by the work authority to isolate a circuit or apparatus when it is needed to provide safe working conditions.	1. An isolation guarantee for work is given to a competent individual who is deemed responsible for obtaining and maintaining a circuit in an isolated and grounded state. 2. The application (usually on a prepared form) for an isolation guarantee on a circuit is made out in advance so that the controlling authority can study it in relation to other applications and can determine the impact on the system (especially for transmission lines). 3. The application can be done verbally when an outage is needed on short notice. Both parties should write down the verbal application on an application form. 4. A joint responsibility for the controlling authority and the work authority ensures that the proper element is identified for an isolation guarantee.
3. Switching is carried out and tags (cards) are placed on the switchgear providing isolation.	1. A switching instruction or switching order form (as seen in Table 2–2)4530 is prepared as a plan for the switching sequence needed to provide the isolation guarantee for the work. 2. If verbal switching instructions are used (radio, etc.), the message should be repeated back for confirmation. 3. After operating any switchgear, the switchgear must be visibly checked to ensure that it is in the correct position. 4. A switchgear that serves as an isolation point for an isolation guarantee for work on a circuit over 750 V must have a visible open point and be tagged.

(continued)

Common Elements	Details
	5. In addition to tagging, reduce risk of an inadvertent operation of a switchgear by using an additional safety measure, where applicable, such as a lock, removing a switch element, removing a loop, disabling a motor-operated disconnect, or opening an extra disconnecting device.
4. The controlling authority certifies and signs a formal document that the circuit is isolated from all sources and will remain isolated until the isolation guarantee is formally surrendered by the work authority.	1. A formal communication, confirming that the isolation guarantee is in effect, is made, often issuing an isolation guarantee number to keep the isolation guarantee distinct from others issued by the controlling authority. 2. Authority to install protective line grounds and go to work is given to the work authority by the controlling authority.
5. The work authority verifies that isolation is complete.	1. Potential checks are carried out to ensure the absence of a normal line voltage before protective line grounds are installed. 2. In some jurisdictions, the protective grounds are controlled and tagged by the controlling authority, while in others the work authority controls the placing of protective line grounds.
6. When work is completed, the work authority surrenders the isolation guarantee for the work.	1. The person holding the isolation guarantee ensures that all workers are clear and that protective line grounds are removed before surrendering the isolation guarantee.

Operating Guidelines

If	Then
Preparing or applying for an isolation guarantee on a circuit . . .	The circuit must be identified by a name, number, or acronym. Any operable device that may be used as a guaranteed device in a lockout/tagout procedure should be labeled in the field and identified on an operating map, an operating one-line, or an operating schematic drawing.
You cross utility boundaries on storm trouble . . .	The lockout/tagout procedure in other utilities may not have the same terms and meaning. For example, a "hold off" tag can be used in one utility to tag an isolating device for an isolation guarantee

(continued)

If	Then
	while it may be used in another utility to block the automatic reclose on a breaker. "Deenergize" can mean isolating a circuit in one utility while in another utility it can mean applying protective grounds on a circuit.
You need a circuit isolated to carry out the work safely . . .	The circuit must be grounded. Working on an isolated circuit using live line methods to avoid installing protective grounds is a trap.
Switching or acting as a switching agent for a controlling operator (dispatch) . . .	Use a written switching order and operating drawing. Get a copy or write down the sequence when received by radio.

A Typical Switching Order

The switching order shown in Table 2–2 was prepared to isolate a set of voltage regulators. When used to isolate a circuit, the circuit nomenclature should be used to describe the exact device to be operated.

TABLE 2–2   Switching Order Form

Switching Order				
Purpose	Isolate Voltage Regulator FR16			
Ordered By	District Control	**Time**		
Completed By	John Lineman	**Time**	10:30	
Sequence Number	*Device Being Operated*	*Operation*	*Tag #*	*Initials*
1	Auto/Manual Switch	Neutral Tap		JL
2	Input & Output Terminals	Voltage Test		JL
3	Auto/Manual Switch	Turn Off		JL
4	Bypass Switches - 3 phases	Close		JL
5	Load Switches - 3 phases	Open		JL
6	Source Switches - 3 phases	Open		JL
7	Source Risers	Remove		JL
8	Load Risers	Remove	NA	JL

Operating Switchgear

3.1 Switchgear Operating Hazards

Hazards	Barriers
Major arcing can result in flashover to grounded parts and molten metal raining down on the operator.	Switchgear must be designed to avoid excessive arcing, and switching orders must be written to support that. The heat from the arc causes the air to become ionized. *Ionization of air* means that the air has become charged with atoms and that the air is changed from being an insulator to being a conductor. Once the air between switch contacts becomes conductive, a follow-through current will flow across the switch gap.

(continued)

3.1 Switchgear Operating Hazards *(continued)*

Hazards	Barriers

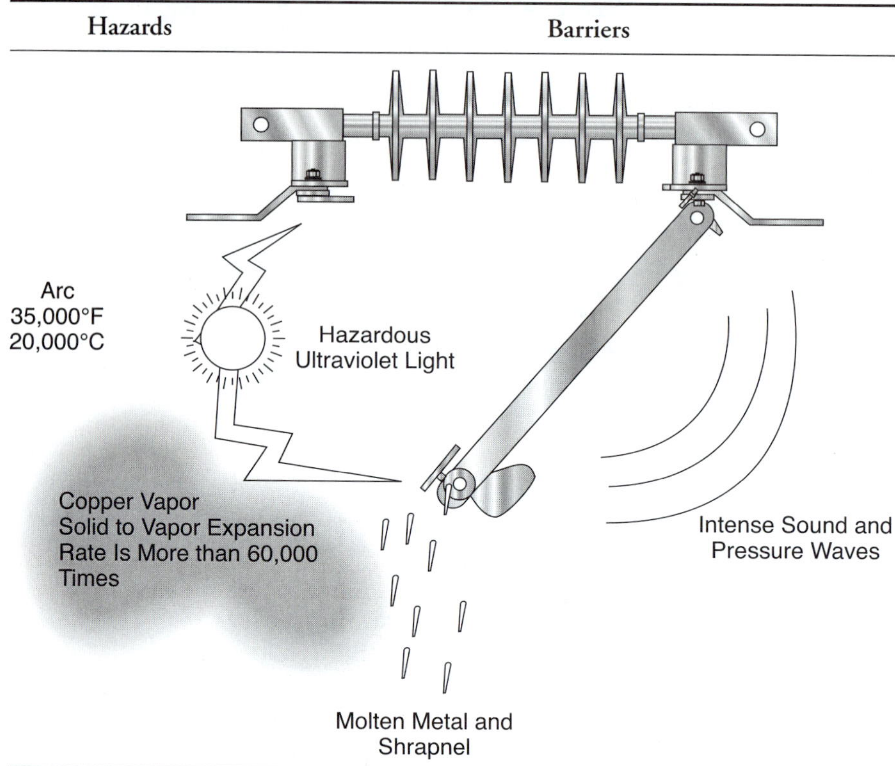

Figure 3–1   The properties of an arc.

If an excessive arc occurs and metal sprays out from the switchgear when switching, some kind of switching error has been made. The properties of an arc are illustrated in Figure 3–1.

A line worker should not be asked to open a device molten unless one of the following conditions is met:
1. The switchgear is designed to interrupt load current.
2. The switchgear is capable of accepting a load-bust tool.
3. The switchgear will not be interrupting any load.

Switchgear can fail with the resultant live parts contacting grounded cases, handles, and so on. The resultant ground fault causes touch and step potentials.

When switching at a device that is operated with a stick, use at least a 10-ft. length of hot stick.

When switching with a handle at ground level, or operating a toggle switch at a breaker, use a ground-gradient mat (see Figure 3–2). Many employers also require rubber gloves.

(continued)

3.1 Switchgear Operating Hazards *(continued)*

Hazards	Barriers

Porcelain pedestal-type insulators have a history of failures.

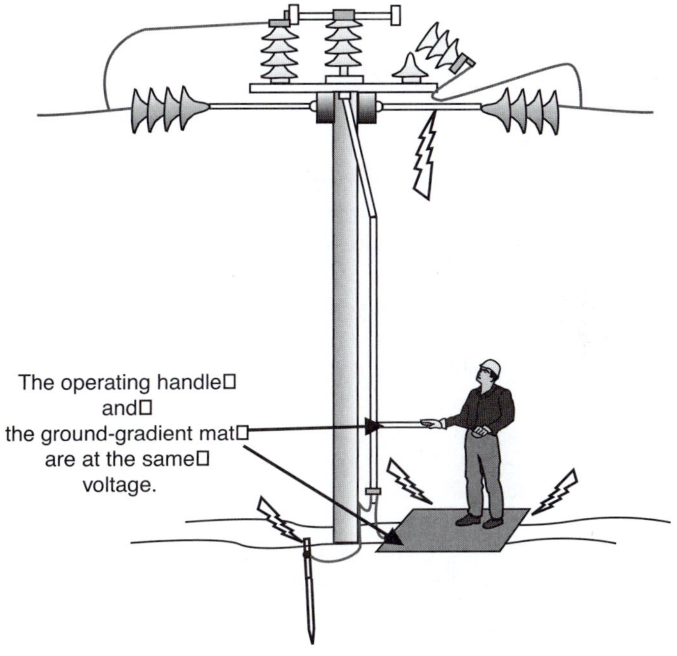

Figure 3–2 Use of a ground-gradient mat.

Hazards	Barriers

Dropping load with switchgear not designed with the capability to interrupt load current will result in uncontrolled arcing.

An ammeter check can be made to confirm that no load is on a circuit before opening a nonload-break disconnect switch. Some employers suggest that a nonload-break disconnect switch be operated by using the *inching method (cracking)*. No arc should be developing as the switch is opened. If an unexpected load exists and an arc starts to develop, close the switch immediately.

If a fairly long length of conductor (no load) is being deenergized by a disconnect switch, the switch should be opened in one fast operation. The employer should specify the allowable length of line that can be dropped for each voltage.

(continued)

3.1 Switchgear Operating Hazards *(continued)*

Hazards	Barriers
When using a telescopic switch stick from the ground, a person can be exposed to a high-potential ground gradient during a failure when a fault current comes down the ground wire.	Keep a distance away from any ground rod at the base of a pole and keep your feet together during the switching operation.
If an open tie switch exists between two different sources the two circuits may not be in-phase. Closing the switch will be like closing in on a short circuit.	To confirm proper phasing between two sources, carry out phasing checks before switching at the open point between two live sources. When the two sources are in-phase, the meter reading on the phasing tester should read near 0 between phase 1 and phase A, and so on. In the field, draw a table (such as the one in Figure 3–3) to record the results of your tests.

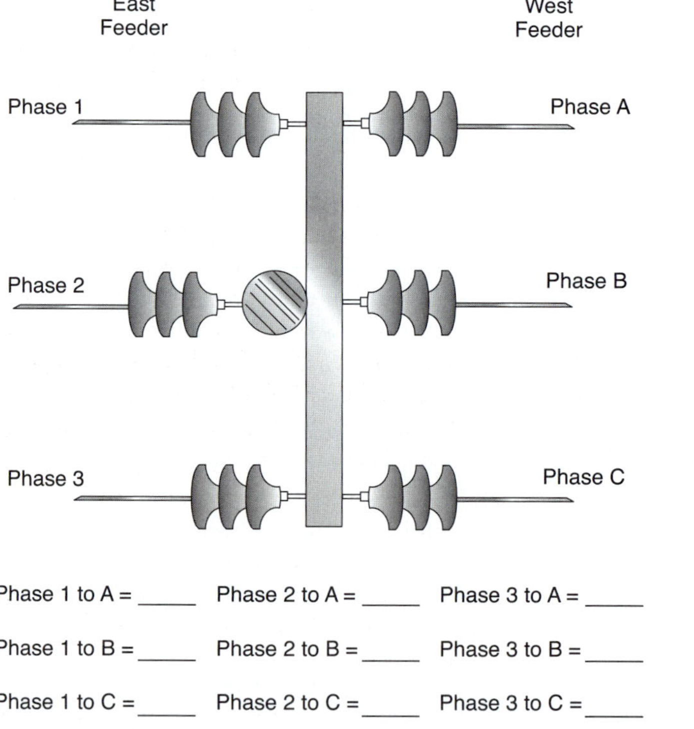

Figure 3–3   Check phasing before switching.

3.2 Overhead Switchgear Capability

Types of Switchgear	Switchgear Capability and Operation
Circuit breaker	*Circuit breakers,* including distribution reclosers and metal-clad breakers, can interrupt fault current and drop load current. Circuit breakers are generally operated remotely from a control room. If there is a reason to open a circuit breaker at the breaker control box, a fault in the breaker can raise the potential on the grounded breaker tank. In a substation, the station grid will keep an operator at the same potential as the tank. In a line setting, an operator should stand on a ground-gradient mat that has been bonded to the tank.
Metal-clad circuit breaker	A *metal-clad breaker* can interrupt fault current and drop load current. It can be operated with a toggle switch or, more commonly, by remote control from a control room. Because of the extremely high-fault current available and the limited space for operation, full face protection and flame-retardant clothing should be worn when racking in or out a metal-clad breaker. A *draw-out circuit breaker* can be in an in-service, test or disconnected position, or it can be completely removed.
Hydraulic recloser	A *hydraulic recloser* can interrupt fault current and drop load current by pulling down the operating lever under the sleet hood. The handle that opens and closes the recloser is often painted yellow. The switch can be closed and held in the closed position to provide a cold-load pickup capability. Holding the switch blocks the instantaneous trip function. The non reclose handle is right beside the open-and-close handle Larger hydraulic reclosers have a low-voltage supply to an electric motor, which allows the heavy contacts to be *opened* and *closed* quickly. Any hydraulic reclosers that can be operated remotely (SCADA) would also have a low-voltage supply to operate the recloser. The recloser can be opened with the handle, but some reclosers must be closed from the control box. The handle must be in the up position, or the remote control or SCADA control cannot close the recloser.

(continued)

3.2 Overhead Switchgear Capability *(continued)*

Types of Switchgear	Switchgear Capability and Operation
Electronic reclosers	Electronic reclosers can interrupt fault current and drop load current. An electronic recloser has a low-voltage supply, which allows the recloser to be operated from a control box. The low-voltage supply to the control box must be in service to close the recloser. The normal method of opening or closing these units is with the operating switch in the control box or remotely through the SCADA system. The main operating switch in a control box can open and close the recloser. Most electronic reclosers have a cold-load pickup feature that is programmed to prevent an instantaneous trip-out when closing after a lockout. An electronic recloser can detect a fairly small phase-to-ground fault. This sensitivity can prevent reenergizing a circuit because the ground fault relay will interpret an unbalanced load as a phase-to-ground fault. Temporarily switching the ground trip toggle switch to the block position will prevent a small ground current from tripping out the circuit. When operating the main switch, an operator should stand on a ground-gradient mat, or other grid, bonded to the control box. The control box can become energized if the recloser malfunctions or the bushings flash over because the control cable bonds together the recloser and control box. Figures 3-4 and 3-5 show two types of electronic recloser control panels.

(continued)

3.2 Overhead Switchgear Capability *(continued)*

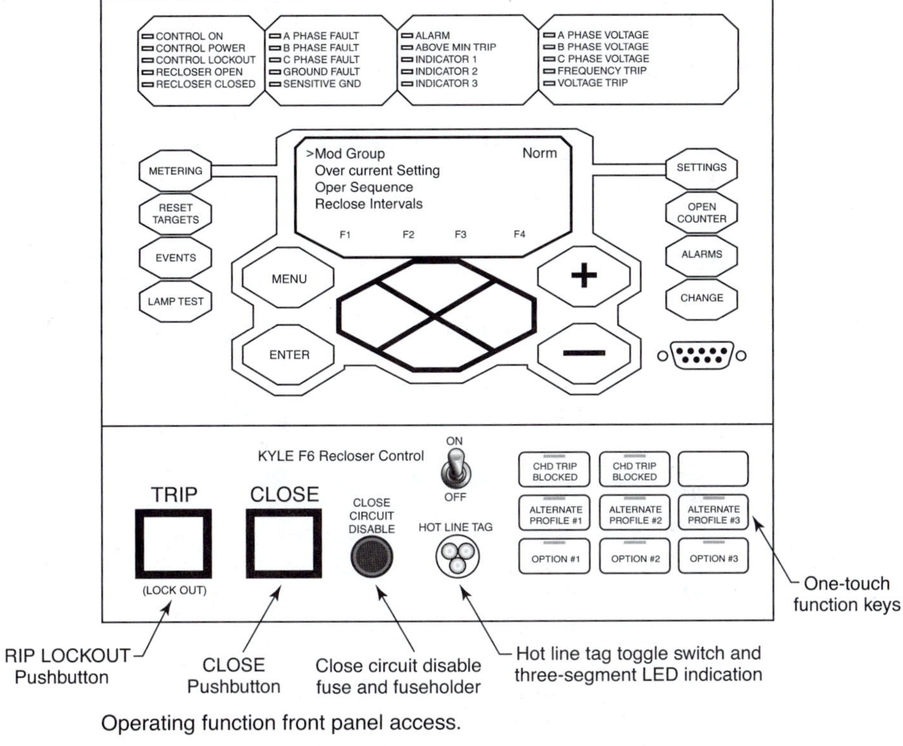

Figure 3–4 Cooper Form 6 control panel.

(continued)

3.2 Overhead Switchgear Capability *(continued)*

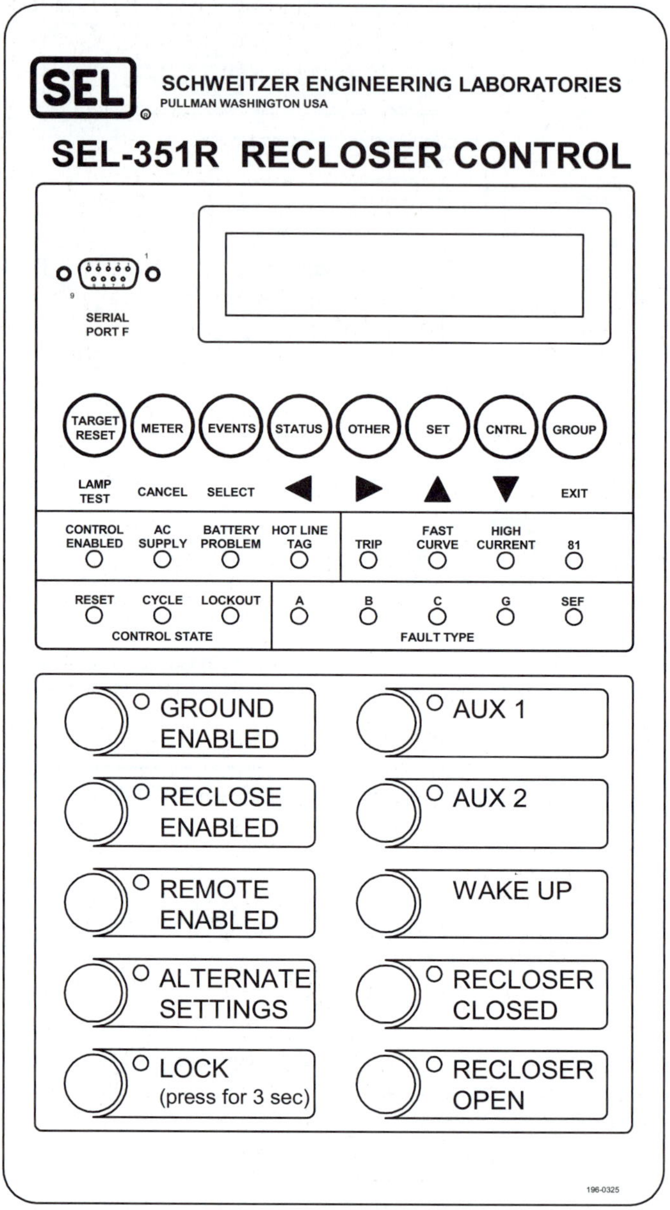

Figure 3–5 SEL-351R recloser control panel.

(continued)

3.2 Overhead Switchgear Capability *(continued)*

Types of Switchgear	Switchgear Capability and Operation
Distribution Automation Switchgear	If switching or working in an area protected by a Distribution Automation (DA) system, then it is critical to communicate your intentions with the system control operators. In this sytem tie switches can close automatically.
Automated switchgear that can communicate back to the control center include SCADA controlled reclosers, remotely controlled circuit breakers and switches such as S&C IntelliTEAM II, S&C Remote Supervisory PME Pad-Mounted Switchgear, and Scada-Mate	Distribution Automation is a computerized system that controls and monitors switchgear beyond the substation fence and it will close a feeder tie switch to feed back upstream to an open point to isolate an upstream permanent faulted section.
Sectionalizer	A *sectionalizer* cannot interrupt a fault current but can be used to drop load current. It is a device that will isolate a faulted section of line while an upstream multishot recloser or circuit breaker is open during a specified second or third interval.
	A *hydraulic sectionalizer* is an oil switch with a mechanism that will open automatically after a fault current goes through the coil a specified number of times. As with reclosers, sectionalizers come with various voltage ratings and current ratings. A hydraulic sectionalizer can be used safely to interrupt load current by pulling down the operating lever under the sleet hood, as with any oil switch. Hydraulic sectionalizers look very much like reclosers because both types are oil switches, but a sectionalizer does not have a nonreclose handle.
Electronic sectionalizer	An *electronic sectionalizer* (see Figure 3–6) cannot interrupt fault current and cannot drop load current unless a load-bust tool is used. An electronic sectionalizer is a slave device that opens during an interval when the upstream multishot device has isolated the circuit.

(continued)

3.2 Overhead Switchgear Capability *(continued)*

Types of Switchgear	Switchgear Capability and Operation

An electronic sectionalizer is like any solid-blade switch: When it is to be operated manually, it requires a portable load-break tool to interrupt load current.

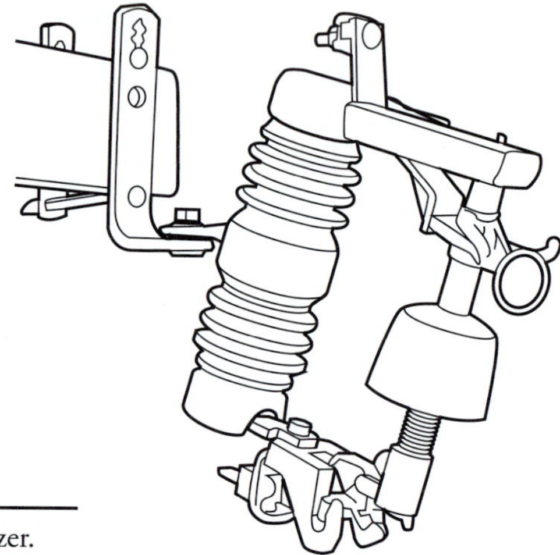

Figure 3–6   Electronic sectionalizer.

Disconnect switch

A *disconnect switch* is any solid-blade switch installed to provide a means to isolate or sectionalize a circuit. Unless a disconnect switch has a load interrupter, it has no capability to interrupt load current or capability to open automatically under fault conditions. Depending on the type, disconnect switches are operated by hot stick or operating handle, or remotely by way of a low-voltage motor.

Figure 3-7 illustrates an in-span disconnect that has no load-break capability unless it is opened with a load-break tool.

This switch may show a rating of 600 amperes, but that means it is capable of carrying 600 amperes, not breaking a 600-ampere load. The switch is operated with a handle at ground level, and the design of the switch installation should include a grid placed below and bonded to the handle. Some utilities require the use of a portable ground-gradient mat.

(continued)

3.2 Overhead Switchgear Capability *(continued)*

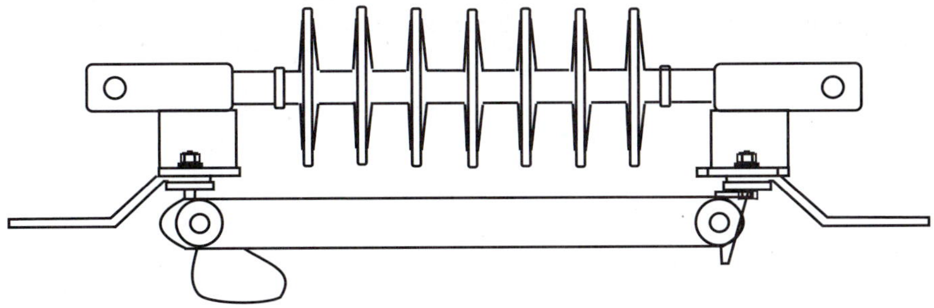

Figure 3–7 Single-phase in-span disconnect switch.

Types of Switchgear	Switchgear Capability and Operation
Load-interrupter switch	A *load-interrupter switch* (also called a *load-break disconnect switch*) is designed to interrupt load current and can be operated without creating a dangerous arc. Such devices can be three-phase gang-operated switches or single-phase switches. Depending on the type, load interrupters are operated by hot stick or operating handle, or remotely by way of a low-voltage motor. Load-interrupter switches can break load and are normally rated at 600 amperes or less. A load-interrupter switch does not open automatically under fault conditions. Figure 3-8 shows a common type of load interrupter. It extinguishes the arc within an interrupter housing by deionizing gases that are exhausted flamelessly through a muffling device.

Figure 3–8 Three-phase load interrupter.

(continued)

3.2 Overhead Switchgear Capability *(continued)*

Types of Switchgear	Switchgear Capability and Operation

Figure 3-9 is a type of load-interrupter switch that is used in substations. The power fuse provides overcurrent protection but cannot drop load current. The load-interrupter switch is used to drop load but cannot interrupt a fault current automatically.

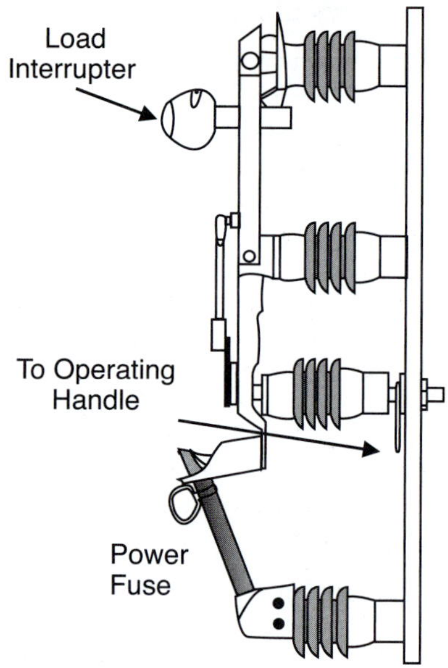

Figure 3–9   One pole of a load-interrupter switch and power fuse.

Fused disconnect/cutout

A *fused cutout* can interrupt a fault automatically (see Figure 3–10), but when the device is opened with a switch stick it has no arc-extinguishing capability other than the air gap between the switch terminals. When opened with a switch stick, a cutout can only interrupt about 15 amperes safely. If closed on a faulted circuit the exhaust gases and molten metal will be expelled and directed down from the vent end of the fuse holder.

At distribution voltages, *single-phase disconnect switches* are normally equipped with hooks to allow for use of a portable load-break tool.

(continued)

3.2 Overhead Switchgear Capability *(continued)*

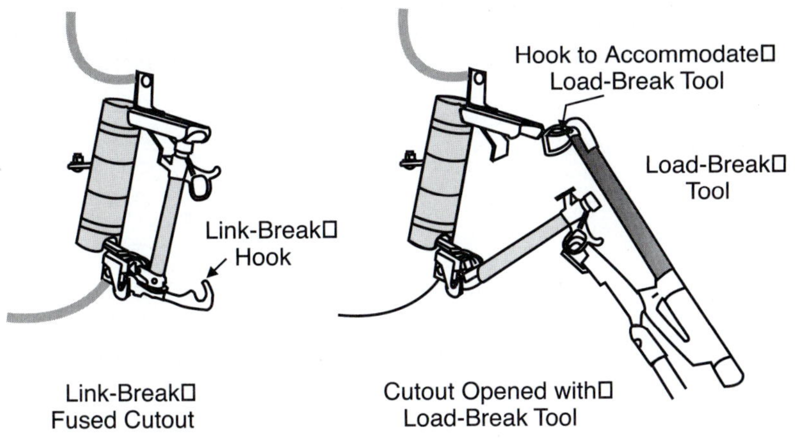

Figure 3–10 Dropping load with a cutout.

Types of Switchgear	Switchgear Capability and Operation
Insulated jumpers and load-break capability	*Jumpers* can be used to break load if equipped with a *load pick-up tool* or if a portable *load-break jumper clamp* (portable switch) is used. Ensure that the correct voltage and load-break rated units are used.
Hot-line clamp	*Hot-line clamps* are not switches, but they are used to conveniently disconnect and connect risers between equipment and circuits. To reduce the risk of an unmanageable arc, a hot-line clamp should not be used to drop any load or even an unloaded transformer winding larger than 10 kVA.

3.3 Underground Distribution Switchgear Capability

Underground switchgear is housed in metal-clad cabinets in substations, in metal pad-mount cabinets, in submersible cabinets, and in vaults/chambers. The reduced clearances to grounded cabinets and objects results in no tolerance for switching errors that result in an arc.

Type	Switchgear Capability and Operation
Protective switchgear that can interrupt a fault and be used to drop load	*Vacuum fault interrupters, SF6 switchgear, metal-clad oil circuit breakers, electronic power fuses* (see Figure 3–11).

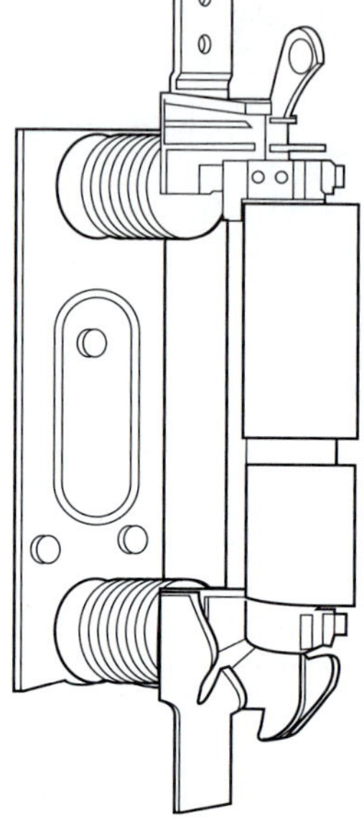

Figure 3–11 S&C electric power fuse.

(continued)

3.3 Underground Distribution Switchgear Capability *(continued)*

Type	Switchgear Capability and Operation
Protective switchgear that can interrupt a fault but cannot be used to drop load	*Solid material power fuses* will interrupt a fault but cannot be used as a disconnect to drop load (see Figure 3–12).

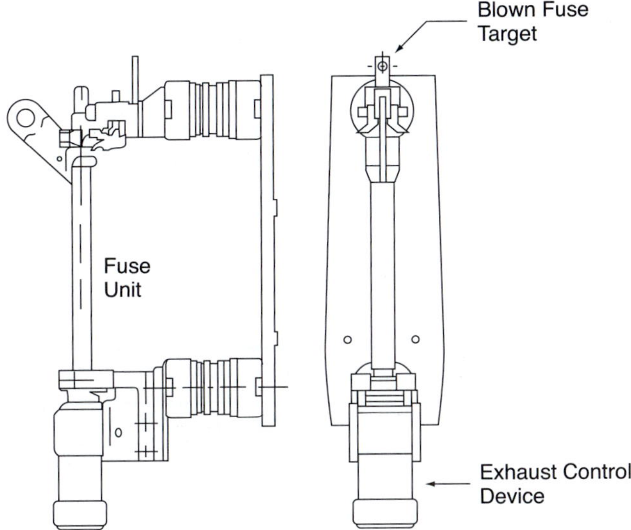

Figure 3–12   Solid-material power fuse. *Source:* Courtesy of S&C Electric Co.

Figure 3-13 shows an elbow in series with an *under-oil bayonet-style fuse* contained in a load-break holder. The fuse provides protection to the transformer by interrupting a fault, but it is not a good device to drop load. To prevent oil from being expelled when the fuse is withdrawn, relieve the pressure in the transformer tank by operating the external relief valve.

(continued)

3.3 Underground Distribution Switchgear Capability *(continued)*

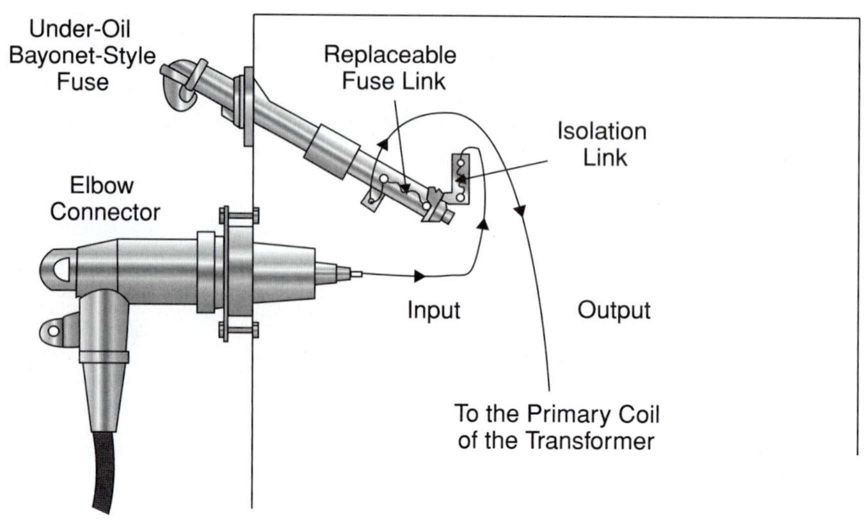

Figure 3–13   Load-break and nonload-break elbows.

Type	Switchgear Capability and Operation
	Three-phase pad-mount transformers often have *dry-well canister fuses,* which are current-limiting fuses that protect transformers mounted in oil-tight dry-well canisters. The current-limiting fuse should not be used to isolate or reenergize a transformer. Even when an isolated condition is assumed, the fuse should be removed with a shotgun stick. Ensure that the proper fuse length and adapters are used because they can be different for different voltages. An incorrect length will not make good contact and will lead to arcing and failure (see Figure 3–14).

(continued)

3.3 Underground Distribution Switchgear Capability *(continued)*

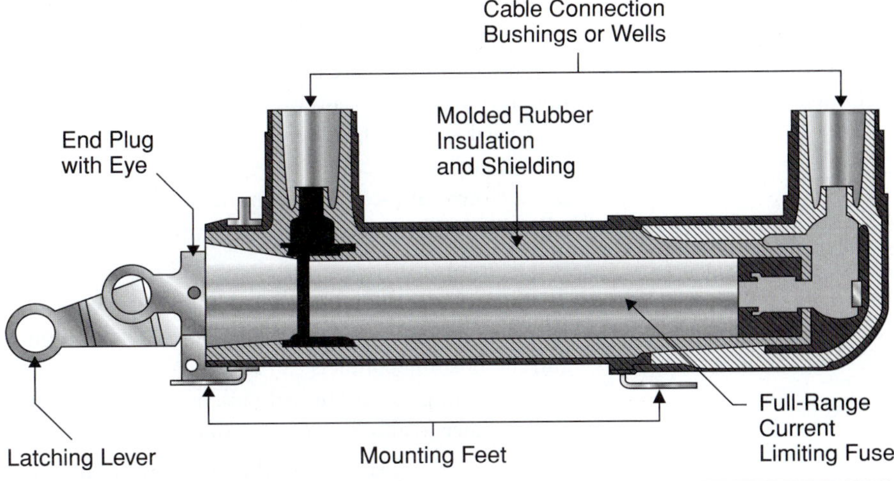

Figure 3–14 A molded dry-well canister.

Type	Switchgear Capability and Operation
Switchgear that cannot interrupt a fault but can be used to drop load	*Nx (nonexpulsion) switchgear* (also called an *arc-strangler switch*) is found in live-front switching cabinets and transformer vaults. It is operated with a live-line tool in the same manner as operating a fused cutout. The Nx switchblade is an arc-quenching load-break device that is available as a current-limiting fuse or as a solid-blade device (see Figure 3–15). The load-break works like a portable load-break tool and must be cocked to operate.

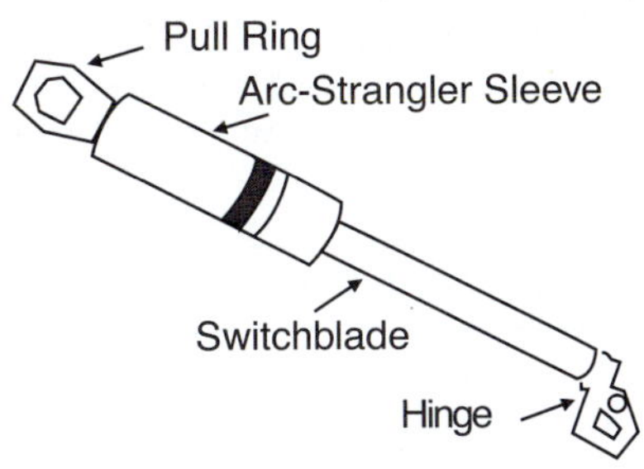

Figure 3–15 An arc-strangler switch.

(continued)

3.3 Underground Distribution Switchgear Capability *(continued)*

Type	Switchgear Capability and Operation
Cable-separable connectors (elbows)	*Separable connectors* are the live-line clamps of an underground system, although fused load-break elbows are available. Load-break elbows are used to isolate equipment or to sectionalize segments of cable. A load-break elbow can be pulled from the load bushing with a live-line tool to interrupt a 200-amp load current. If the lubricant between the elbow and the bushing has dried, an elbow does not always come all the way off with the first pull. An arc between the load-break pin and the bushing contacts can spill over to the outside of the bushing to a grounded tank or cabinet. Use an elbow puller to allow removal in one motion. Some nonload-break elbows (Figure 3-16) should not be operated alive. A nonload-break elbow can have a current rating as high as 600 amps and is often used to terminate a main feeder into a switching cabinet.

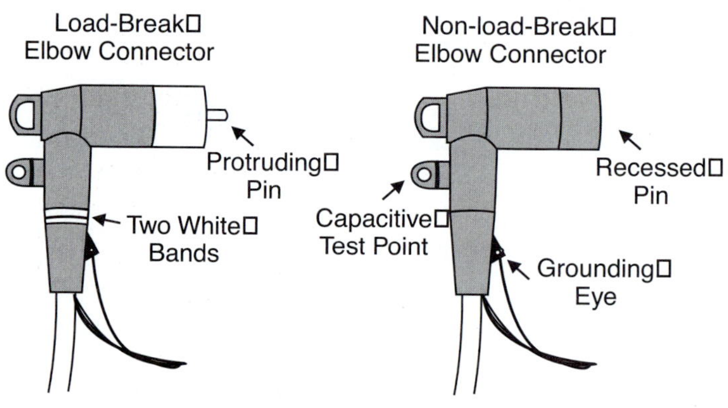

Figure 3–16   Load-break and nonload-break elbows.

3.4 Understand the Reason Switchgear Is Being Operated

If	Then
A length of line, with a load on it, is to be isolated or interrupted	A load current will be interrupted and an arc will have to be extinguished. Switchgears, such as a circuit breaker, a recloser, or a load interrupter disconnect switch, are capable of dropping load and quenching an arc. A disconnect switch without a load interrupter is not designed to interrupt any load. It may have a rating of 600 amps, but that refers to the load it can carry through the switch continuously.
A length of line, with no load on it, is to be isolated	The line can be isolated with an ordinary disconnect switch. The length of line that can be dropped with a disconnect switch depends on the voltage. For example, an air break switch, without a load interrupter, can, with permission, drop up to 3 miles (5 km) of 230 kV or up to 16 miles (26 km) of 50 kV.
Switchgear is opened to break parallel in a circuit that is being fed from two directions	Some current and voltage will be interrupted. If the switch breaks parallel between different stations or two unequal lengths of line, a voltage difference (recovery voltage) will occur and a current flow will be interrupted upon opening the switch. A control room operator has the information needed to calculate the amount of current to be interrupted. Switchgears capable of interrupting load current can break parallel easier than can an ordinary disconnect switch.
Switchgear is closed to make parallel between two sources	There can be a voltage difference and current flow created when the switch is closed. Unless there is a restrike, all switches should be capable of picking up the rated load current. Closing switchgears between two circuits fed from two different substations that are fed from two different transmission systems can be dangerous because a large voltage difference may create a large current flow through the switch. A system control operator should have control of such switchgears.

(continued)

3.4 Understand the Reason Switchgear Is Being Operated

(continued)

If	Then
A loop must be opened or closed (opening a circuit fed from two directions but fed from the same circuit breaker breaks a loop)	Unless a loop is excessively long, it can be closed or opened with an ordinary disconnect switch or a hot-line clamp. An example of a loop is where a new line is constructed next to an old one, such as at a road-widening project. The new circuit can be energized in parallel with the old circuit by tying it in at both ends and forming a loop. Any load on the old circuit must be opened before breaking the loop and isolating the old circuit.
A switch is to be closed, as a test, to determine if a line is still faulted	Switchgears can be closed without an arc. A major arc will occur if a switch is closed in on a fault and the powerline worker immediately reopens the switch. It is important to let the normal protective switchgear trip out the circuit automatically.

3.5 Load-Bust/Break Tool

A load-bust tool can be hooked up to a fuse cutout or disconnect switch to provide a parallel path for load current through the tube of the load-break tool. The initial downward pull on the tool charges an internal spring. At a certain point in the downward pull, the spring is released, resulting in a high-speed separation of the contacts. Any arc is extinguished inside the chamber by the fast elongation of the arc and the release of deionizing gases formed from the surrounding chamber material. A load-bust tool is a bypass for the load current. The opening of the circuit and any resultant arc occur inside the load-break tool.

S&C Catalog Number Shown on Tool	Rating	
	Maximum Switch Rating	Interrupting Capacity
5300R3	27 kV	600 A
4700R2	15 kV	200 A
5400R3*	38 kV	600 A

*The 5400R3 must not be used on switches rated at less than 15 kV because the fuse tube or blade travel would be too short to accommodate the operating stroke of the load buster.

Caution: When the bottom of a switch is alive, insufficient clearance from the tool to the neutral may create a hazard.

3.6 Current-Limiting Fuses

Types	Purpose
Partial-range current-limiting fuse (An ordinary expulsion fuse is not current limiting. It will limit the duration of an arc but not the magnitude. A current-limiting fuse introduces a high resistance after the fuse element melts.)	A *partial-range current-limiting fuse* (also called a *backup current-limiting fuse*) is used in series with a fused cutout to protect equipment from the high energy levels available in a high-fault-current location. An *air-insulated, partial-range current-limiting fuse* is installed right on top of a fused cutout (see Figure 3-17) and is available in different sizes and speeds to coordinate with a cutout fuse link. For example, a 25K current-limiting fuse coordinates with a 25K cutout fuse link.

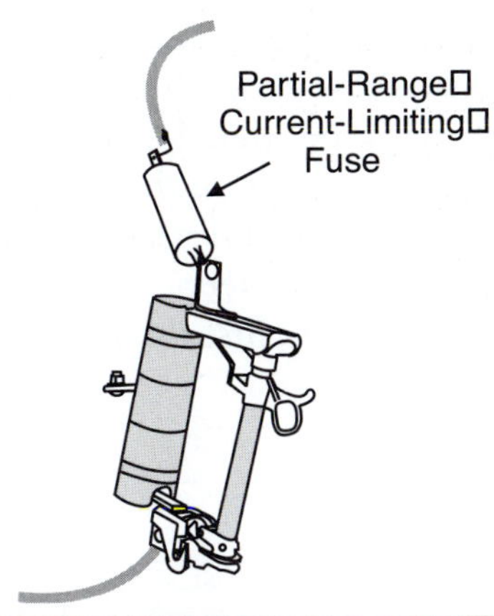

Figure 3–17 Application of a partial-range current-limiting fuse.

(continued)

3.6 Current-Limiting Fuses *(continued)*

Types	Purpose
	An *under-oil submersible partial-range current-limiting fuse* is used in series with low-current protective devices such as a bayonet-style under-oil expulsion fuse, a cartridge fuse, or a fuse link.
	If a *transformer cutout fuse* is found open, the current-limiting fuse—if properly coordinated—should not be damaged. Testing or replacing the current-limiting fuse before energizing the transformer with the cutout will reduce the risk of an explosive failure if the transformer is faulted.
A full-range current-limiting fuse	A *full-range current-limiting fuse* (also called a *general-purpose current-limiting fuse*) is most often used to protect underground or metal-clad equipment. It is not used in series with another fuse but is the only fuse needed to protect the equipment. A full-range current-limiting fuse can be air-insulated and used for high-amperage-rated applications in live-front switchgear or as an under-oil submersible current-limiting fuse (bayonet-style mounted fuse) used with dead-front equipment.

3.7 Nonreclose Feature on Breakers and Reclosers

Putting a breaker or recloser in a nonreclose position gives a crew working on a circuit the assurance that the circuit will not be reenergized automatically if an incident on the job trips out the circuit.

The nonreclose position does not guarantee that a circuit will trip out during an accidental contact, or that it will trip out quicker. It only ensures that a circuit will stay out of service after it is tripped out.

Place a tag at the breaker or recloser to prevent other personnel from closing the switchgear without first checking with the crew working on the circuit.

CHAPTER 4

Trouble Investigation: No Power

4.1 Trouble Investigation of an Individual Customer's Service

Check voltage at . . .	Action
The customer's breaker	If the voltage is normal at the top of the main breaker, look for an open or faulty main breaker or look for blown fuses (breakers) in the panel. Some utilities will not check beyond the meter.
The customer's meter base	If the voltage is normal at the top of the meter base, then check for a faulty meter or bad connections in the meter base.
The transformer feeding the customer	If the primary circuit is supplying normal power to the transformer, check for a blown primary fuse, loose connections, bad neutral connections, a blown surge arrester, an open current-limiting fuse, or a faulty transformer.

4.2 Trouble Investigation of an Over-Current Situation

During an over-current situation, a line crew is called out after circuit breakers, reclosers, sectionalizers, or fuses have gone through their sequences and have isolated any circuit with a permanent fault.

If	Then
There has been wind, ice, wet snow, or lightning.	The problem is often the traditional broken conductors, broken poles, or fallen trees and branches.
Protection tripped out on a nice clear day.	The fault is sometimes due to public contact, such as a car accident, tree cutting, or a boom contact. An outage on a clear day suggests that a patrol should be carried out before energizing the circuit.
An overload is the likely cause of an outage.	The most common fix is phase balancing. The planning engineer often assumes that the load on the feeder is balanced among the three phases. When the load is not balanced, one phase of the three-phase system is carrying more than its share of the load and the protective device will trip out.
The circuit has been out for a while, especially during peak-load periods.	There is a high initial inrush current when the switch is closed, and the line trips out again. A heavily loaded circuit may have to be picked up a section at a time.

4.3 Trouble Investigation of a No-Power Call on an Overhead System

Common Problems	Restoration
Lightning is the most frequent cause of a transient fault. At the flashover point, a high-voltage arc establishes a path of ionized air to ground. A high follow-through current is established through the ionized air and causes an over-current fault.	In an urban area, a patrol on the circuit is always wise before closing in the circuit. In a rural setting, lightning without high winds should allow reenergization without a patrol.

(continued)

Common Problems	Restoration
When *wind, ice,* or *wet snow* have been present, the cause of an outage is often a phase-to-ground fault. A phase-to-ground fault is the cause of about 70% of permanent faults. Probable causes are a bad insulator, tree contact, broken conductor, or animal contact. Accidental contact by the public includes car accident, crane contact, sailboat contact, ladder contact, or antenna contact.	In a populated area, patrol the line before reenergization. In a rural setting, patrol the line unless special circumstances indicate that local knowledge would allow reenergization.
On *very cold or very hot days,* suspect an overload problem. Over-current due to an overload occurs when the customer demand exceeds the specified setting of the circuit protection. Circuit protection does not differentiate between over-current due to an overload or a fault.	The most common fix to an overload problem is phase balancing. When the load is not balanced between phases, one phase of the three-phase system is carrying more than its share of the load and the protective device will trip out. A low-resistance tree contact on a tap protected by a fuse can be seen as an additional load by the upstream three-phase device.
On an underground cable, a phase-to-ground fault can occur due to an *insulation breakdown, dig in, or a driven fence post.*	Other than an overload, a blown fuse on an underground cable normally calls for checking out the cable before reenergizing.
The circuit has been out for awhile, especially during peak-load periods. There is a high initial in-rush current when the switch is closed and the line trips out again.	A heavily loaded circuit may have to be picked up one section at a time. On an electronic recloser, the handle can be held closed momentarily until the in-rush current drops. On a hydraulic recloser, holding the handle closed will not prevent a trip-out.

(continued)

Common Problems	Restoration
After patrolling, there is *no apparent cause* for a permanent line outage, but the protective device trips out each time the line is reenergized.	• The cause may be a faulty surge arrestor. • There may be a fault downstream past the section that was patrolled. When the downstream protective devices do not trip out in proper sequence, it could be that incorrect fuses were installed during previous work. • A punctured dead-end insulator could also be the cause.

4.4 Trouble Investigation of a No-Power Call on an Underground System

Situation	Action
Individual customer	A no-power call from an individual customer starts with a voltmeter check at the meter base. For no voltage, check out other customers fed from the same transformer. The voltage checks will show if the problems are a faulted service cable or a transformer problem.
System at large with fault indicators	If fault indicators have been installed (as shown in the relatively simple system illustrated in Figure 4–1), the line crew would find the switch at the riser pole open, check the fault indicators at each transformer, and find the fault indicators at the first three transformers showing that the fault passed through there and that the fault did not go through the indicator at transformer 126. The conclusion would be that the fault lies somewhere between transformers 125 and 126.

(continued)

Situation	Action

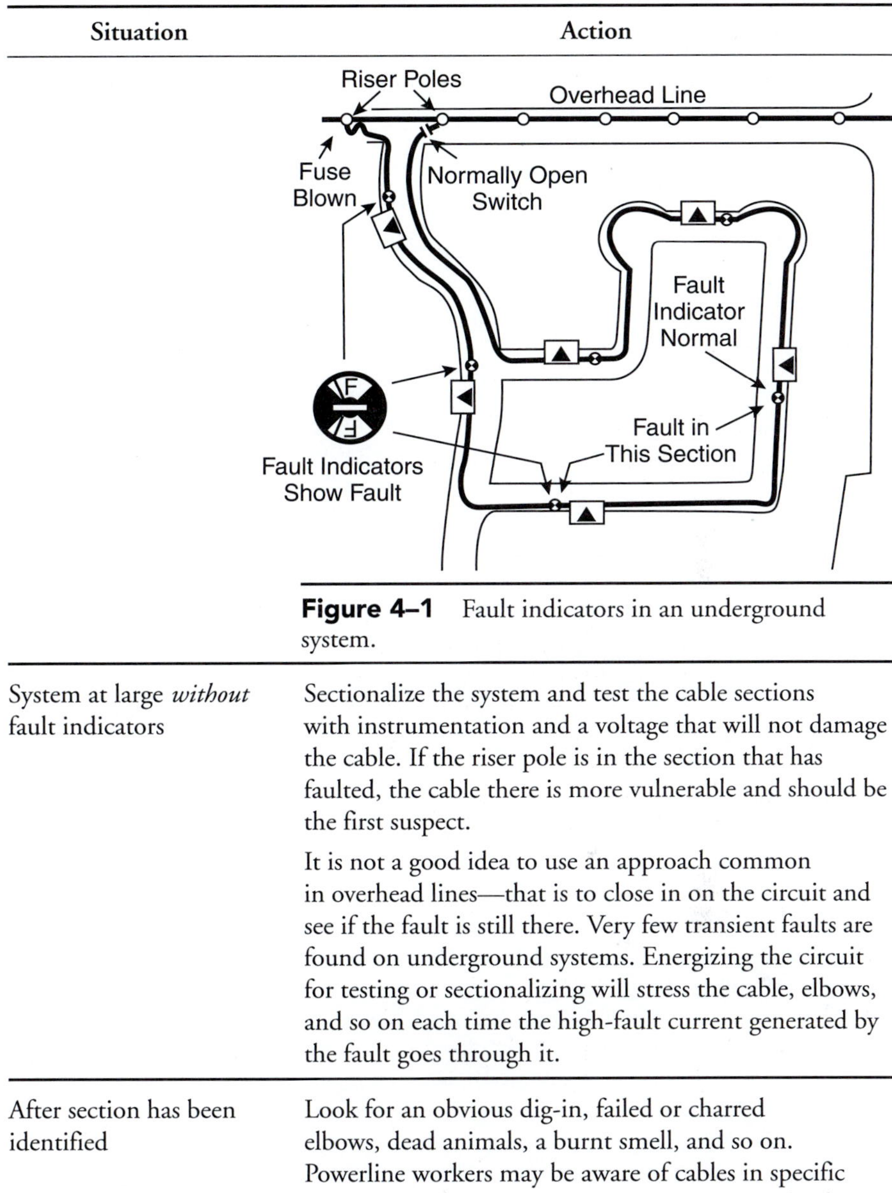

Figure 4–1   Fault indicators in an underground system.

Situation	Action
System at large *without* fault indicators	Sectionalize the system and test the cable sections with instrumentation and a voltage that will not damage the cable. If the riser pole is in the section that has faulted, the cable there is more vulnerable and should be the first suspect. It is not a good idea to use an approach common in overhead lines—that is to close in on the circuit and see if the fault is still there. Very few transient faults are found on underground systems. Energizing the circuit for testing or sectionalizing will stress the cable, elbows, and so on each time the high-fault current generated by the fault goes through it.
After section has been identified	Look for an obvious dig-in, failed or charred elbows, dead animals, a burnt smell, and so on. Powerline workers may be aware of cables in specific locations that are vulnerable due to aging, overloading, improper backfill, stresses from settling soil, and so on. Use instrumentation, such as a DC Hipot, to locate a faulty section of cable.

CHAPTER 5

Trouble Investigation— Voltage Problems

5.1 Voltage Standard

If	Then
A customer is complaining about high or low voltage.	Measure the voltage at the meter base.
	A utility has to meet a voltage standard when supplying power to a service. Table 5–1 shows the voltage ranges that must be met and are measured at the point of delivery, usually at the meter base. Standard Range A refers to a favorable voltage level. Standard Range B is tolerable for a short term but should be reported so that corrective action is initiated. Once voltage falls outside of range B, the customer equipment will not operate properly.

(continued)

If	Then
The voltage appears to be acceptable but the customer continues to complain.	Install a recording (data logger) voltage monitoring instrument.

TABLE 5–1   Voltage Standards

Service Voltage	Range A Minimum	Range A Maximum	Range B Minimum	Range B Maximum
% of Nominal	95%	105%	91.7%	105.8%
120/240 3-wire	114/228	126/252	110/220	127/254
240/120 4-wire	228/114	252/126	220/110	254/127
208Y/120 4-wire	197/114	218/126	191/110	220/127
480Y/277 4-wire	456/263	504/291	440/254	508/293

5.2 Trouble Investigation: Low Voltage

Finding the Cause of Low Voltage

Step	Action	Details
1	Check the voltage at the customer meter base.	If the voltage is low, check for any recent increase in the load a customer is drawing. An increase in load could make the length or size of secondary conductors inadequate for the additional load. The planning engineer or engineering standards books have voltage-regulation charts for secondary bus and services.
2	Remove all load from the transformer and take a voltage reading at the transformer.	If the voltage reading is normal at the unloaded transformer, the low-voltage problem must be due to the length and/or size of the secondary or possibly an overloaded transformer. A recording voltmeter may have to be installed if the problem is intermittent. If the voltage reading is low at the unloaded transformer, the primary voltage is low and the cause of the problem is upstream. Other customers should be affected by upstream problems. A quick

(continued)

Finding the Cause of Low Voltage *(continued)*

Step	Action	Details
		fix to an individual customer is to make a change to the transformer tap, if equipped.
3	Check out any upstream voltage regulator or transformer with a load tap changer (LTC).	A voltage regulator could be stuck in a low position. Put the regulator in "manual" position and raise the voltage. Similarly, the LTC at the substation transformer could be malfunctioning.
4	Check out any capacitors on the line.	Switched capacitors should be in service during daily peak-load times and switched out of service during lightly loaded times. Check to ensure that the time clock or other control is working. Check that the motor-operated oil switches are closed during peak-load periods.
5	On a heavily loaded circuit, check the phase balance.	The planning engineer may show that the circuit can carry the load, but if one phase is carrying more than its share, it could be overloaded and cause an excessive voltage drop on that phase.

5.3 Trouble Investigation: High Voltage

Finding the Cause of High Voltage

Step	Action	Details
1	Check the voltage at the customer's meter base.	If the voltage is high and the customer is close to a distribution substation or close to a voltage regulator, the voltage is often relatively high at these locations to provide good voltage farther downstream. Check the voltage of a neighbor fed from a different transformer to find out if the high voltage is unique to the one customer that is complaining.

(continued)

Finding the Cause of High Voltage *(continued)*

Step	Action	Details
2	If the neighbors have normal voltage, do a ratio test on the transformer causing the high voltage.	A transformer can occasionally suffer some shorted-out turns in the coil and cause high voltage to customers.
3	Check out the substation load tap changer (LTC) or upstream regulator.	A voltage regulator could be stuck in a high position. Put the regulator in "manual" position and lower the voltage. Similarly, the LTC at the substation transformer could be malfunctioning.
4	Check out any nearby capacitors.	Capacitors that have not automatically switched out of service when the circuit is lightly loaded can raise the voltage during off-peak times. Check to ensure that the controller (load controller, VAR controller, time clock, etc.) of the capacitor switch is working. In other words, check that the motor-operated oil switches are open during off-peak load periods.
5	If all other conditions are normal, change the voltage taps only on the transformers feeding customers with the high voltage.	Depending on the manufacturer, each tap will raise or lower the secondary voltage by 4.5 percent or 2.5 percent. The nameplate on the transformer must be consulted. *The transformer must be isolated before changing the voltage tap.* The feeder voltage will change during the day and at different seasons. Therefore, changing the taps at the transformer could produce extreme voltages if the primary voltage returns to a normal level.

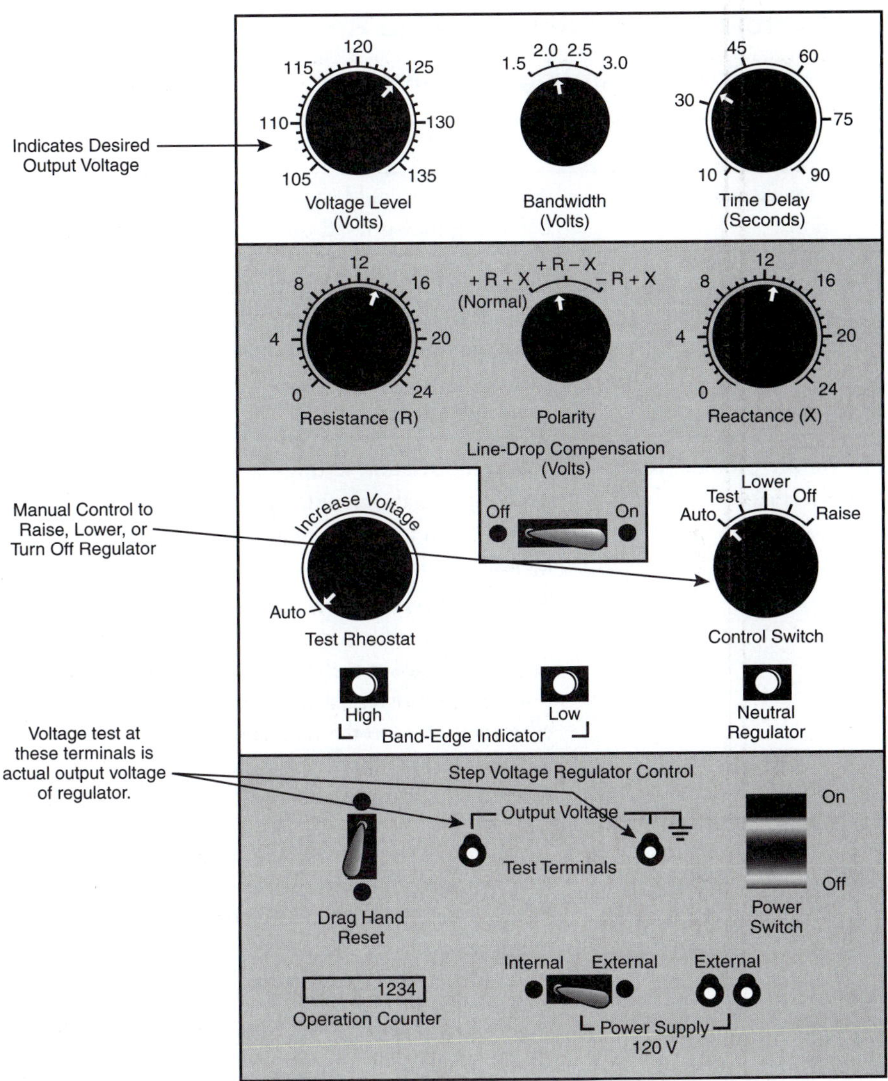

Figure 5–1 Regulator control.

5.4 Trouble Investigation: Voltage Regulator
Guide for Troubleshooting a Regulator

If	Then
There are excessive regulator operations.	The regulator could be overloaded. Phase balancing may unload the affected regulator.
There is a voltage complaint on a circuit where there is a regulator.	Often, the tap-changing mechanism gets stuck on a tap. Raise and lower the voltage using the auto–manual switch, and then return the switch to automatic to see if the tap changer will move of its own accord.
The voltage problem is not due to a sticking tap changer.	The compensation settings may be out of date due to changes to the circuit, such as the installation of a capacitor, an upstream regulator, new conductors, or an increase in load. As a *temporary* measure, adjust the voltage using the auto–manual switch and then turn off the switch. Test the voltage at the customer and at the voltage test studs in the regulator control box. Now customers close to the regulator and/or at the end of the line will be exposed to more extreme voltages when the load current changes.
The regulator is chattering or hunting. The regulator is changing taps continuously.	The bandwidth or time-delay settings are incorrect. As a *temporary* measure, adjust the voltage using the auto–manual switch as before. This fix is temporary because customers may be exposed to extreme voltages when the load current changes.
The regulator does not respond to the previous solutions.	The regulator may be damaged. Take the regulator out of service. If the tap changer cannot be put in the 0 or neutral position, arrange to take the circuit out of service by opening the source switch with a load-break tool, then open load switches and close the bypass.
The regulator is at the maximum boost position but is not capable of supplying the acceptable voltage.	1. An increased load in the circuit may have reduced the input voltage to the regulator. 2. Phase balancing could unload the affected phase. 3. An upstream or downstream regulator may be needed.

(continued)

Guide for Troubleshooting a Regulator *(continued)*

If	Then
The actual position of the 0 or neutral tap is not certain.	The pointer on the tap position indicator may be broken and/or the neutral indicating light is not working. Sometimes the specifications for the control settings do not have the 0 tap at the center. If in doubt, take the regulator out of service by opening the source switch with a load-break tool, as before.

5.4.1 Operating a Regulator

If	Then
The source and load voltages are not equal when the bypass is closed.	The *regulator* will be subjected to a short circuit. The most critical operation involving a regulator is ensuring that the source and load voltages are equal when the bypass is about to be closed.
The input and output voltage is to be zeroed.	The *control switch* is used to manually raise or lower the regulator to the neutral position. The *neutral indicator* light should come on when the regulator is in neutral position. To equalize the voltage, the auto–manual switch in the control box is turned to manual operation. The voltage can then be raised or lowered until it reaches the 0 or neutral tap. When the neutral tap is reached, the switch is turned to the off position. The bypass switch (see Figure 5–2) can then be closed, and the input and output cutouts can be opened to isolate the regulator.

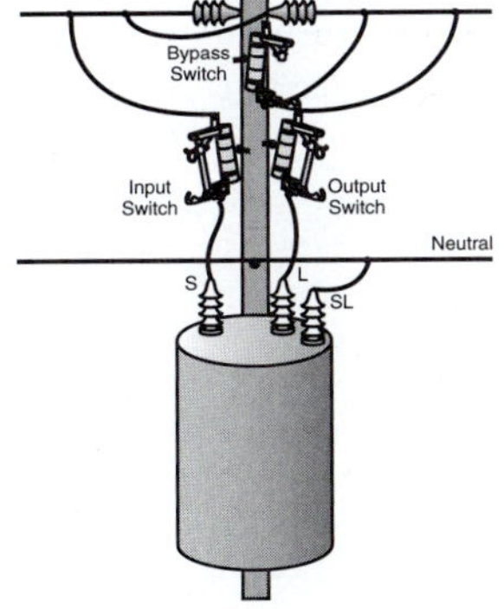

Figure 5–2 Typical regulator switching arrangement.

(continued)

5.4.1 Operating a Regulator *(continued)*

If	Then
A circuit is temporarily fed in reverse through a regulator.	The regulator will continue to adjust the voltage on its load side, which is now the source. The voltage sensor measures the input voltage instead of the output voltage. Typically, the regulator goes to the maximum boost or to the maximum buck position. The regulator should be bypassed and removed from service before the circuit is fed in reverse.

5.5 Trouble Investigation: Capacitor Bank

If	Then
Working on a low-voltage complaint, especially during times when there is a high customer demand.	The capacitor should be in-service. The timer or the low-voltage supply to the oil switches may be out of service. If available, the clock should be checked to ensure it is set so that the capacitors are boosting the voltage during peak load time.
Working on a high-voltage complaint, especially during times when there is low customer demand.	The capacitor should be out of service. The timer or the low-voltage supply to the oil switches may be out of service. If available, the clock should be checked to ensure it is set so that the capacitors are removed from service to prevent high voltage during lightly loaded times.
The tank is bulged or leaking.	The capacitor is defective.
Capacitors are in-service but may not be working.	Use an ammeter to check the amperage on the risers feeding the capacitors. The current a capacitor bank draws can be measured or calculated. The following are some typical current readings for a 25-kV system: 600 kVAR bank 13.9A 450 kVAR bank 10.4A 300 kVAR bank 7.0A 225 kVAR bank 5.2A

(continued)

5.5 Trouble Investigation: Capacitor Bank *(continued)*

If	Then
	Use the following equation to calculate the expected current for other voltages on wye systems: $$I = \frac{kVAR\ of\ Capacitor\ Bank}{3 \times Phase\text{-}to\ Ground\ kV}$$ where I = current in one phase kVAR = rating (of capacitor bank) kV = phase-to-ground voltage
A capacitor bank is to be removed from service.	Wait at least 5 minutes to place grounds to let the discharge resistors drain the voltage to a level below 50 volts (see Figure 5–3.) **Figure 5–3** Capacitor construction with discharge resistor. Anytime a unit is exposed to contact by people, the terminals of a capacitor should always be left shorted out, even when on a storage dock.
A capacitor is returned to service.	Wait 5 minutes after isolation to allow the voltage on the capacitor to drain before reenergizing.
Energizing capacitors with a fuse cutout.	The line voltage across the cutout can be double if proper contact is not made the first time and an immediate second attempt is made to close. If possible, energize capacitors using an oil switch.

5.6 Trouble Investigation: Voltage Flicker

If	Then
Blinking lights or intermittent shrinking of a television picture, especially in windy conditions.	Look for a loose neutral or other bad connections as possible sources of flickering lights, especially during windy or heavy-load conditions.
There are blinking lights or intermittent shrinking of a television picture, on a regular basis, daily.	Large motors, arc welders, X-ray machines, and electrical arc furnaces have loads with varying demands, are mostly unbalanced, and have a poor power factor. Under starting conditions, these loads draw considerably more current than when operating. A voltage flicker is noticed by other customers served by the same feeder. A typical voltage flicker standard is represented in the graph in Figure 5–4. For example, it can be seen from the graph that a 3% voltage dip five times per minute would be objectionable. Report voltage-flicker problems to the planning engineer.

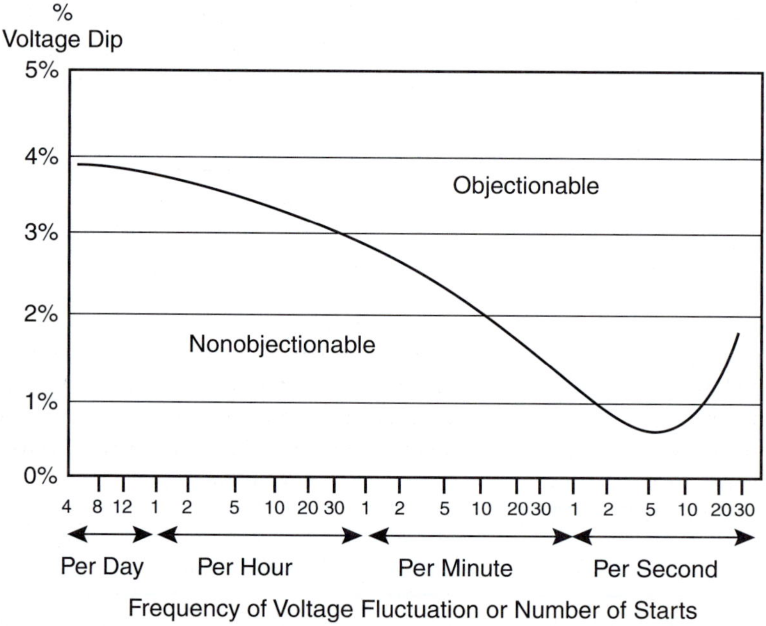

Figure 5–4 Threshold of objectionable voltage.

5.7 Trouble Investigation: Ferroresonance

When resonance occurs, the voltage can build up from two to nine times the normal phase-to-ground voltage. The increase in voltage damages equipment, which is evidenced by rumbling and whining noises at transformers, arcing at insulators, and sparkover at arresters. Ferroresonance can be avoided by removing one of the causes of the phenomenon. By changing the design of the installation or changing the switching procedure, the capacitance, inductance, series connection, or loading can be changed.

Prevention of Ferroresonance

Causal Factor	Action
Operating single-pole switches to energize or deenergize an underground cable feeding a three-phase transformer allows the transformer coils to be an inductive load in series with the circuit.	Install a three-phase gang-operated switch.
If there is no load on the transformers when they are energized, the low resistance in the circuit causes a higher voltage when resonance occurs.	Keep a load on the transformer when it is being energized or deenergized. The increased resistance in the circuit will lower the effects of resonance.
A three-phase bank with a grounded wye primary shorts out the two windings that are part of the series circuit, causing induction.	At the planning stage, where possible, use a transformer with a wye primary and the wye point connected to the neutral.
Changing the length of the underground cable feeding the three-phase transformer will change the amount of capacitive reactance in the circuit.	Changing the length of the cable as a retrofit is probably an expensive option. Calculating the length needed to avoid ferroresonance is complex. The changing inductive reactance that occurs before the transformer core becomes saturated means that there is also a range of capacitive reactance that will at some point be equal to the inductive reactance.

(continued)

Prevention of Ferroresonance *(continued)*

Causal Factor	Action
A wye primary with a floating (ungrounded) neutral is susceptible to ferroresonance.	Temporarily ground the floating neutral of the wye primary, which will short out the series circuit. A three-phase bank with a grounded wye primary shorts out the two windings that are part of the series circuit causing induction.

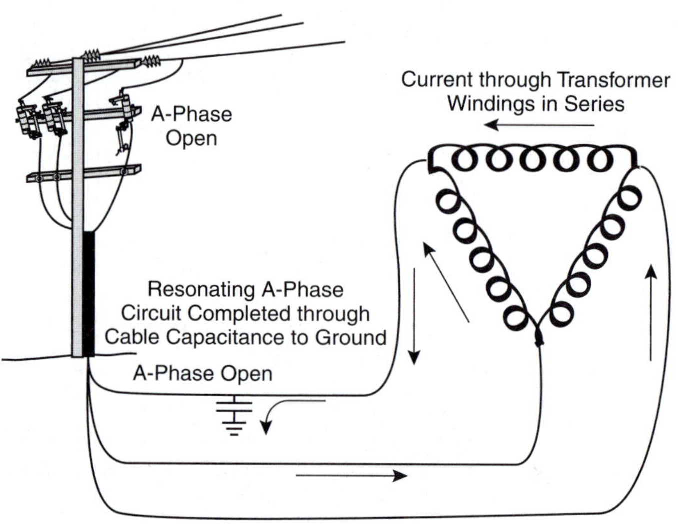

Figure 5–5   An example of ferroresonance.

5.8 Trouble Investigation: Tingle Voltage

Tingle voltage occurs when there is an unacceptable voltage between the neutral and earth. Figure 5–6 shows how a circuit can be established through a person between a shower control and a main drain. Connect an AC/DC voltmeter between the neutral and remote earth (a temporary ground rod about 50 feet [15 m] away). The voltmeter should be capable of measuring as low as 0.1 volt. Under some conditions, a cow can feel a potential difference as low as 0.5 volt. In a shower, a person can feel a potential difference as low as 2 volts.

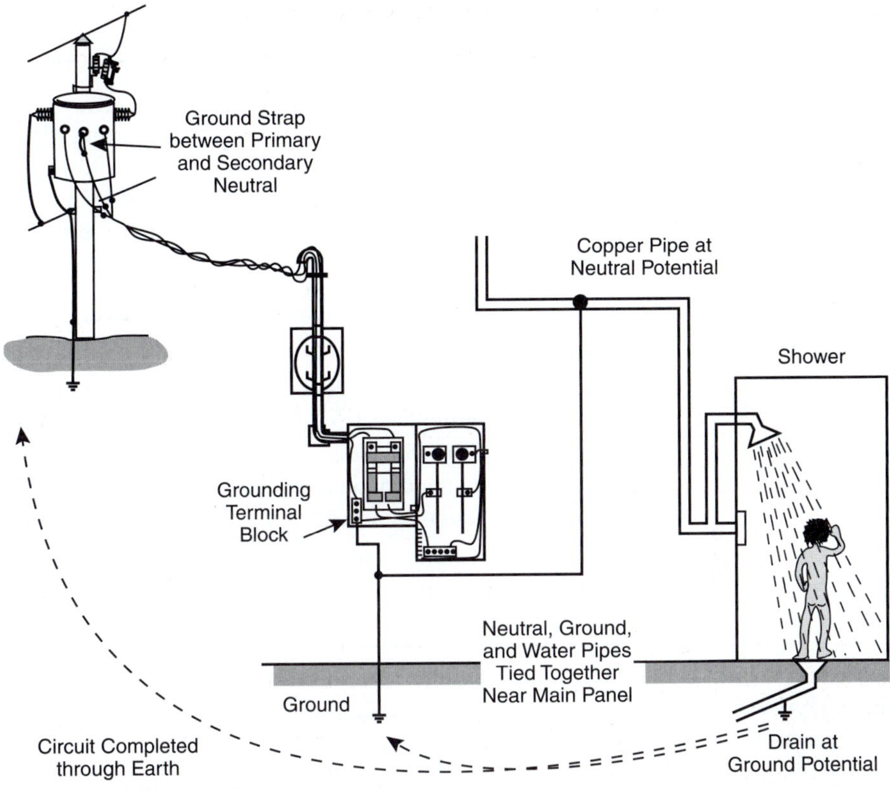

Figure 5–6 Tingle voltage in a shower.

Finding the Source for Tingle Voltage

If	Then
A DC voltage is found between the neutral and earth.	The likely cause is communications circuits or cathodic protection on a nearby pipeline.
An AC voltage is found between the neutral and earth.	Disconnect the customer at the transformer but leave the neutral connected. If the voltage drops to 0, the problem originates with customer equipment. If the voltage remains the same, it is a neutral-to-earth voltage problem. Ensure that the neutral is intact all the way back to the substation.

(continued)

Finding the Source for Tingle Voltage *(continued)*

If	Then
The common neutral is intact back to the substation.	Start sectionalizing the primary feeder, starting with the primary circuit downstream from the customer. Continue to sectionalize until there is a drop in neutral-to-earth voltage at the complaining customer. A drop in voltage will mean the section of line causing the problem has been found.
The section causing the neutral-to-earth voltage is found.	Check for a defective neutral. Check for the existence of a large customer on the same circuit. Check for defective equipment causing a ground current by disconnecting the customer and looking for a drop in neutral-to-earth voltage at the complaining customer. If the customer is on a heavily loaded single-phase circuit, an unacceptable voltage buildup may be on the neutral if the neutral is poorly grounded.
The source of the complaint is a high system neutral voltage and there is a need for an engineering solution.	Engineering solutions include the following: • Additional ground rods are driven to lower the voltage on the neutral. • A larger neutral conductor can be strung to promote more current flow through the neutral and less through earth. • A conversion to three phase will promote more current returning to the source in the other phases and less current flow in the neutral.
An immediate solution is required.	Have the customer install a tingle voltage filter. In an emergency, with concurrence from the supervisor, split the secondary neutral from the primary at the transformer. Install a warning sign on the pole to warn others about the split neutral.

5.8.1 Hazard: Neutral Separation at the Transformer

In extreme cases, a high neutral voltage on the utility supply system will be isolated from the customer's neutral by removing the connection between the primary neutral and the secondary neutral at the transformer. The separation of the neutral (see Figure 5–7) is a hazard to utility personnel because there could be a potential difference across the open circuit between the primary and secondary neutral. Under fault conditions, the voltage on the transformer tank or on the down-ground could be lethal.

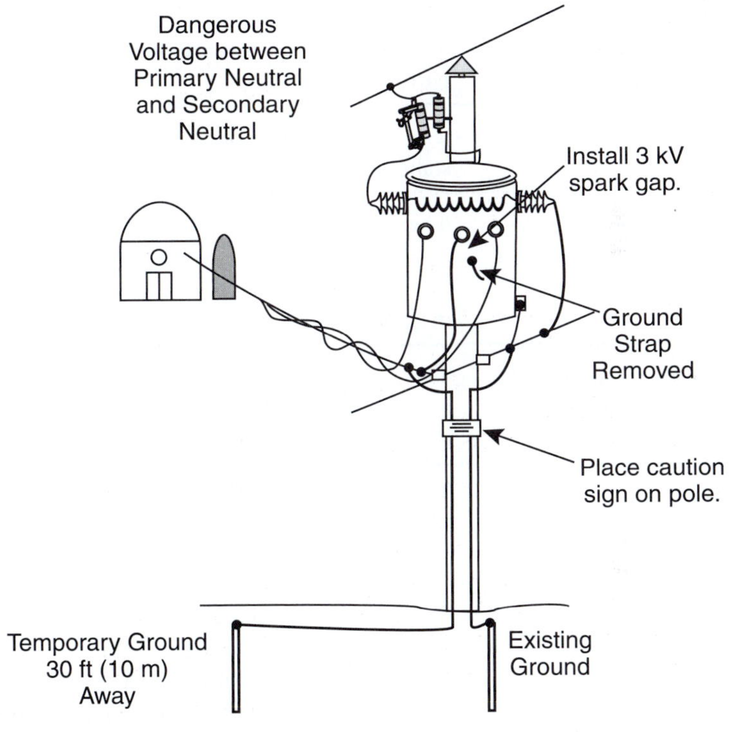

Figure 5–7 Neutral separation at transformer.

A gas-tube type of surge protector can be installed between the secondary neutral and the transformer tank. During a voltage surge, the protector sparks over internally and temporarily connects the secondary neutral to the transformer tank. A caution sign is normally placed on the pole to alert powerline workers to the hazard of the separated neutrals.

5.9 Trouble Investigation: Radio and Television Interference

The source of radio and television interference (TVI) is not always obvious and is not always due to the electrical-utility facilities. Having a technician who uses specialized instrumentation is the best way to find difficult TVI sources.

Finding the Source of TVI

If	Then
The interference is intermittent.	The probable cause of intermittent noise is loose hardware.
The interference is continuous.	The probable cause of continuous noise is from a stable source such as a defective insulator.
The interference appears to be weather dependent.	The noise is most likely from the powerline and not from within the customer's home if the interference occurs during dry weather but disappears during a rain.
The interference is prevalent throughout the neighborhood.	The TVI has a strong source, and the powerline is a probable cause when the noise is prevalent throughout the neighborhood.
The interference affects only one customer.	The noise probably comes from equipment in the building when a weak source affects only one customer.
The interference occurs at similar specific times.	When the noise occurs during specific times, such as during working hours, the TVI could be from industrial equipment such as an arc welder. This type of source should also produce a noise throughout the neighborhood.
The interference on the customer's television occurs on both the video and the audio.	The source is strong when both the audio and the video are affected. The audio signal of a television is an FM signal and is not as vulnerable to interference. If only the audio is affected, the problem is probably within the customer's equipment.

(continued)

Finding the Source of TVI *(continued)*

If	Then
The interference is from devices in the customer's premises.	Devices in the home that have been known to cause noise are an electric motor, a fluorescent light, electronic equipment, a doorbell transformer, flashing decorative lighting, an aquarium pump, a heating pad, a dimmer switch, or a refrigerator butter conditioner.
The interference is from an outdoor source other than a powerline.	Check for potential sources such as radio and television transmitters or two-way radio transmitters used by police, utilities, and others.
The noise is the same across all television channels.	Depending on the strength of the noise, channels 2 to 6 are affected first, channels 7 to 13 are affected next, and UHF is almost never affected. Suspect the television itself if the noise is constant for all channels.

CHAPTER 6

Transformer Connections and Trouble Investigation

6.1 Check Out a Transformer

Depending on the circumstances, checks and tests should be done to ensure a transformer is ready for service.

Checks and Tests	Action
Check the nameplate transformer(s) for and voltage rating.	Check the nameplate for the correct voltages. The nameplate in Figure 6–1 shows that the transformer is an oil-filled, three-phase unit to be fed from a 27.6/16-kV primary. The secondary will feed a three-phase 600/347-V service. Both the primary and the secondary are wye connected. The transformer has a tap changer with 2.5% increments.

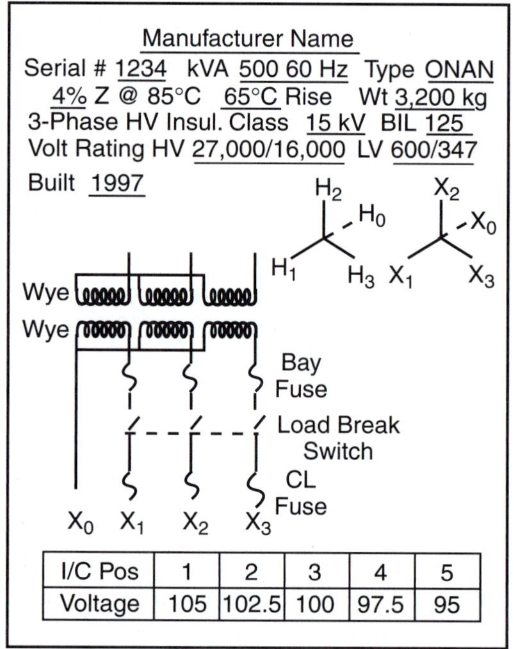

I/C Pos	1	2	3	4	5
Voltage	105	102.5	100	97.5	95

Figure 6–1   Nameplate for a three-phase transformer.

For three single-phase transformers banked together, the supplied secondary voltage will be dependent on the following:
- The voltage rating of the secondary coil.

(continued)

6.1 Check Out a Transformer *(continued)*

Checks and Tests	Action
	• Whether the transformer secondaries are interconnected in a wye or delta configuration. • Whether the secondary coils inside the transformer are connected together in series or in parallel.
Test for transformer turns ratio. (A ratio test will ensure that no short circuits exist between turns in the windings.)	Energize the *high-voltage* coil with a low voltage, such as 120 volts. Measure the input voltage and the output voltage with a voltmeter. (see Figure 6–2). The ratio is calculated as: $$Input \div Output = Transformer\ Ratio$$ **Caution:** Transformers work both ways! For this test, the low-input voltage must be connected to the *high*-voltage winding. If the input voltage is connected to the low-voltage winding, the transformer will be a step-up transformer and a lethal voltage will appear at the high-voltage terminal. Apply 120 Volts AC H_1 H_2 V **Figure 6–2** Turns ratio field test.
Set transformer taps.	Some transformers have off-load tap changers. When changing the position of the tap changer, additional or fewer turns are applied to the primary winding. Depending on the manufacturer, each tap raises or lowers the secondary voltage by 4.5% or 2.5%. The nameplate shown in Figure 6–3 has a 2.5% tap.

(continued)

6.1 Check Out a Transformer *(continued)*

Checks and Tests	Action

Taps	
%	
105	A
102.5	B
100	C
97.5	D
95	E

Figure 6–3   A nameplate showing tap settings.

An external switch knob can be turned to the various positions to change the taps. *The transformer must be isolated before turning the knob.* On some older transformers, the cover must be removed for access to the tap-changer switch knob.

If equipped with tap changers, each transformer used for a three-phase service must be on the same voltage tap.

Impedance: Check the nameplate for its impedance before banking it together with other transformers.

The impedance of the transformers in the bank should be within 0.2% of each other to avoid having the transformer with the lowest impedance take a greater share of the load. In other words, if one transformer has an impedance of 2%, the impedance of the other transformer should be between 1.8% and 2.2%.

Insulation test.

A 1,000-V insulation tester (megger) can be used to test a transformer for damaged insulation. Remove any ground from the X_2 terminal for the test.

The readings between the high-voltage terminal and the low-voltage (X_1, X_2, or X_3) terminals should be infinite. Readings between the low-voltage terminals and the transformer tank should be infinite. The insulation test readings with a three-phase transformer should be like those shown in Table 6–1.

(continued)

6.1 Check Out a Transformer *(continued)*

Checks and Tests	Action

TABLE 6–1   Insulation Resistance Tests on a Three-Phase Transformer

HV to LV Insulation Resistance		LV Insulation Resistance to Tank	
Test Connections	*Readings*	*Test Connections*	*Readings*
H_1 to X_1, X_2, or X_3	Infinite (∞)	X_1 to Tank	Infinite (∞)
H_2 to X_1, X_2, or X_3	Infinite (∞)	X_2 to Tank	Infinite (∞)
H_3 to X_1, X_2, or X_3	Infinite (∞)	X_3 to Tank	Infinite (∞)

Continuity test.	A 1,000-volt insulation tester (megger) can be used to test a transformer for coil continuity. Remove any ground from the X_2 terminal for the test.
	Readings between each of the primary terminals and each of the secondary terminals should be 0, as shown in Table 6–2.

TABLE 6–2   Continuity Tests on Three-Phase Transformer Windings

Continuity Tests of HV Windings		Continuity Tests of LV Windings	
Test Connections	*Readings*	*Test Connections*	*Readings*
H_1 to H_2	0	X_1 to X_2	0
H_2 to H_3	0	X_2 to X_3	0
H_3 to H_1	0	X_3 to X_1	0

Testing for load current.	The actual load on a transformer can be measured and calculated in the field by measuring V_1 and V_2 and I_1 and I_2, as shown in Figure 6–4 and applying the measurement to the following formula.

$$\frac{(I_1 \times V_1) + (I_2 \times V_2)}{1,000} = total\ kVA\ load$$

(continued)

6.1 Check Out a Transformer *(continued)*

Checks and Tests	Action

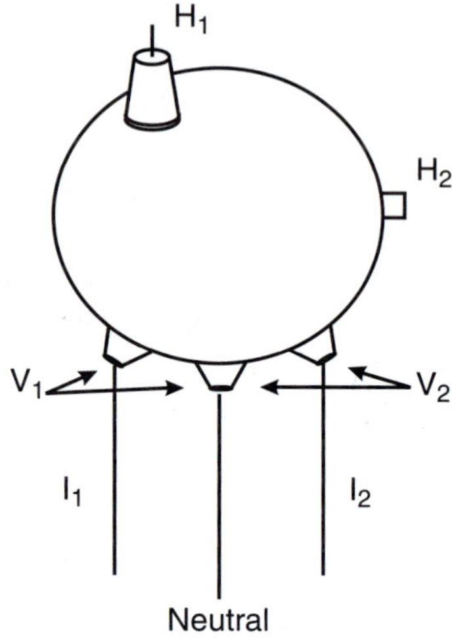

Figure 6–4 Calculating the actual transformer load.

If a fairly high degree of accuracy is needed, *then* the current and voltage on each leg should be taken at the same time to reduce the effects of a load shift from one side to the other.

Suitability for delta or wye primary systems.	The voltage rating of a transformer primary coil must be compatible with the applicable circuit. The voltage impressed across the primary coil will depend on whether the coil is connected in a wye (phase-to-neutral) or a delta (phase-to-phase) configuration.
	Delta: The transformer must have two insulated high-voltage terminals for a delta connection.
	Wye: Transformers intended for connection to wye systems can be constructed with only one insulated high-voltage terminal.
Always check the voltage.	After the installation of a transformer, it is essential to do a secondary voltage check as a final test to ensure that the transformer output is correct.

6.2 Three-Phase Transformer Connections

Type	Connections
Wye–wye **supplies** 120/208 volts 240/416 volts 277/480 volts 347/600 volts	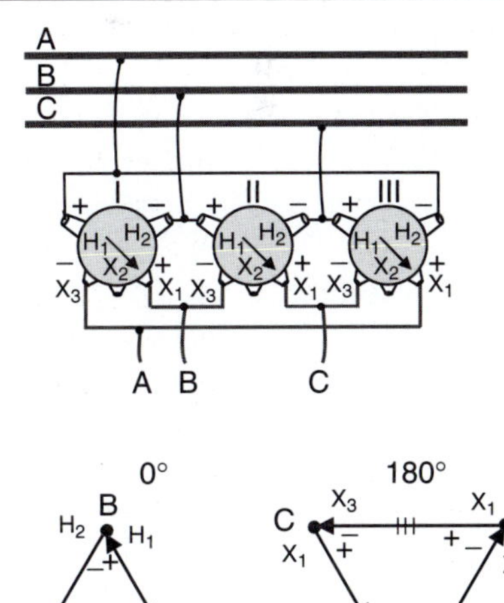

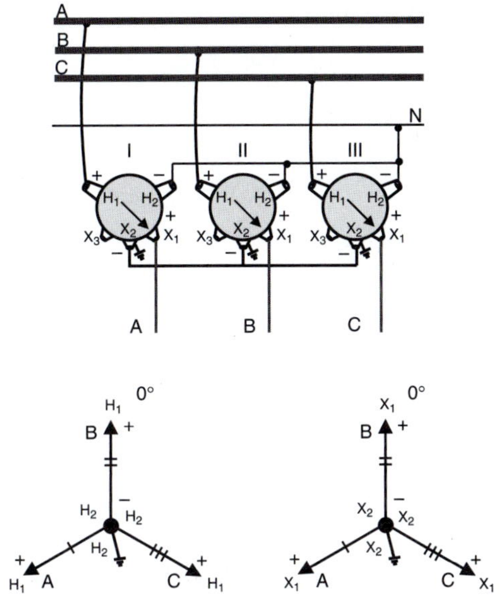

Figure 6–5   A wye–wye transformer bank.

Note: The primary neutral must be connected to the secondary neutral.

| **Delta–delta supplies**
120 volts
240 volts
480 volts
600 volts | |

Figure 6–6   A delta–delta transformer bank.

(continued)

6.2 Three-Phase Transformer Connections *(continued)*

Type	Connections
Wye–delta supplies 120 volts 240 volts 480 volts 600 volts	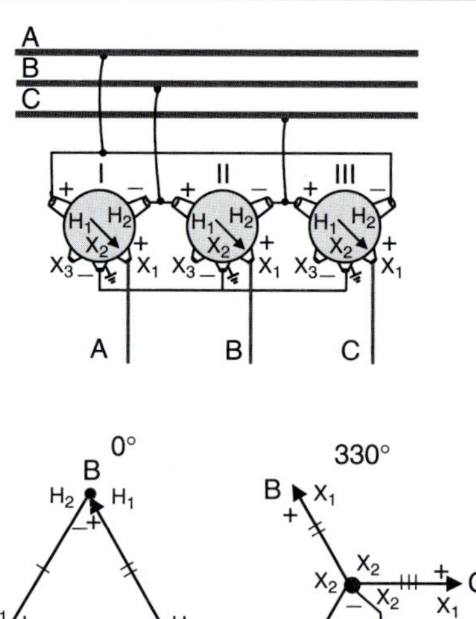

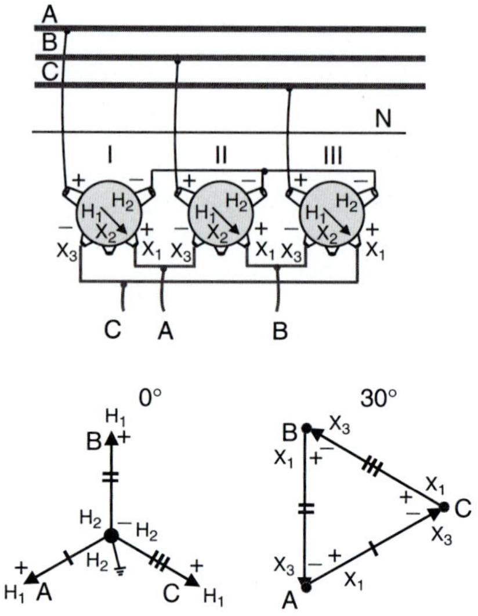

Figure 6–7   A wye–delta transformer bank.

Note: The primary neutral must be left floating (ungrounded).

| **Delta–wye supplies**

120/208 volts

240/416 volts

277/480 volts

347/600 volts | |

Figure 6–8   A delta–wye transformer bank.

Note: The secondary neutral must be well grounded.

6.3 Three-Phase Secondary Connections

6.3.1 Standard North American Three-Phase Voltages

Table 6–3 shows three-phase voltages available from common distribution transformer secondaries in North America.

The three-phase voltages available to a customer are dependent on the standard voltage in that country times $\sqrt{3}$ or 1.732. For example, the following are common three-phase voltages:

North America: $120 \times 1.732 = 208V$
Europe: $220 \times 1.732 = 380V$
Other countries: $240 \times 1.732 = 416V$

TABLE 6–3   **Standard North American Three-Phase Voltages**

Transformer Secondary	Type of Three-Phase Service	Coil Arrangements Inside the Tank	External Configuration
120/240	120/208	Parallel	Wye
	240/416	Series	Wye
	120	Parallel	Delta
	240	Series	Delta
240/480	240/416	Parallel	Wye
	240	Parallel	Delta
	480	Series	Delta
277	277/480	NA	Wye
347	347/600	NA	Wye
600	600	NA	Delta

6.3.2 Internal Secondary Connections

After choosing the correct transformer(s), the secondary voltage from a three-phase transformer bank will depend on the following:

- Whether the transformer secondary is interconnected as wye or delta.

- Whether a transformer with a center-tapped secondary coil has the two parts of the coil inside the tank arranged in parallel or in series. When the two secondary coils are interconnected in series inside the tank, the secondary voltage is double the voltage of the two coils connected in parallel (see Figure 6–9).

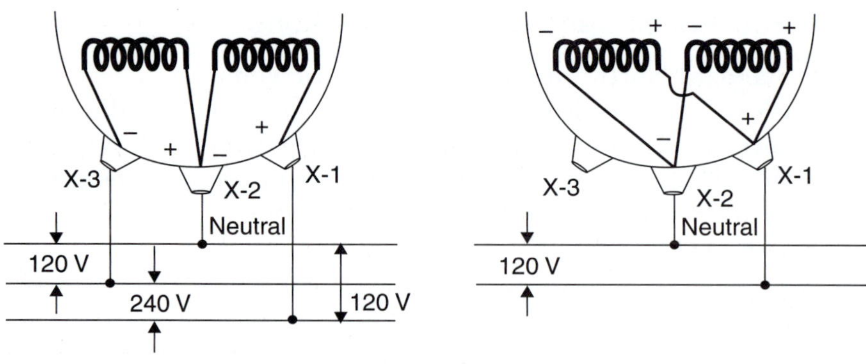

Figure 6–9 *Series and parallel secondary connections.*

6.3.3 Three-Phase Secondary Connection Details

Service Type	Connection Details
Wye 120/208V with 120/240V lighting service. Center transformer: • Secondary coil in series • Often larger	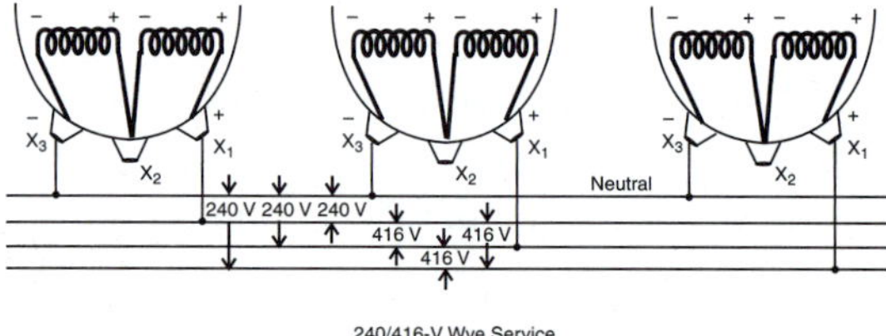 120/208-V Wye with 120/240 Lighting Service **Figure 6–10** A three-phase 120/208V service.
Wye 240/416V service from 120/240V transformers. • Secondary coils are in series. • Any ground strap on the center bushing is removed.	240/416-V Wye Service **Figure 6–11** A three-phase 240/416V service.

(continued)

6.3.3 Three-Phase Secondary Connection Details *(continued)*

Service Type	Connection Details
Wye 240/416V service from 240/480V transformers. • Secondary coils are in parallel. • Center bushing is grounded.	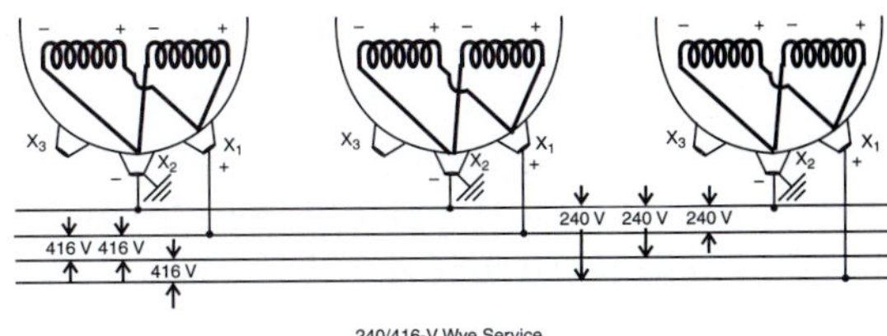

Figure 6–12   A three-phase 240/416V service.

| Delta 240V with 120/240V lighting service.

• One phase from this service will be 210V to ground (wild phase). | 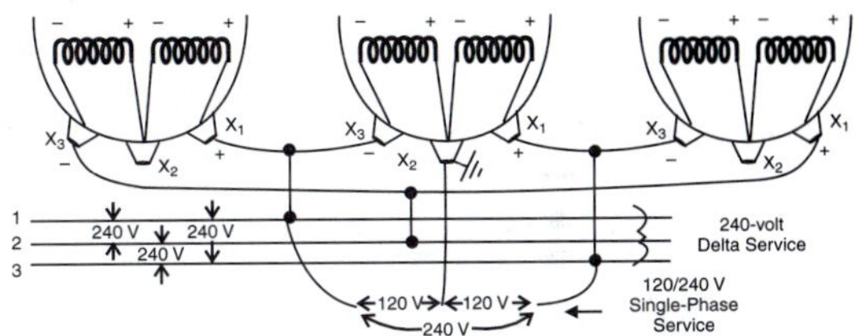

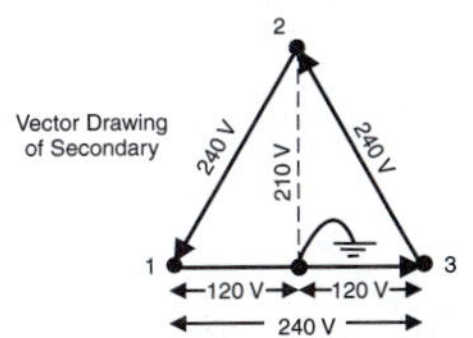 |

Figure 6–13   A three-phase 240V service.

| Wye 347/600V or 277/480V service has similar connections.

• Voltage based on 347V or 277V transformer secondary. | 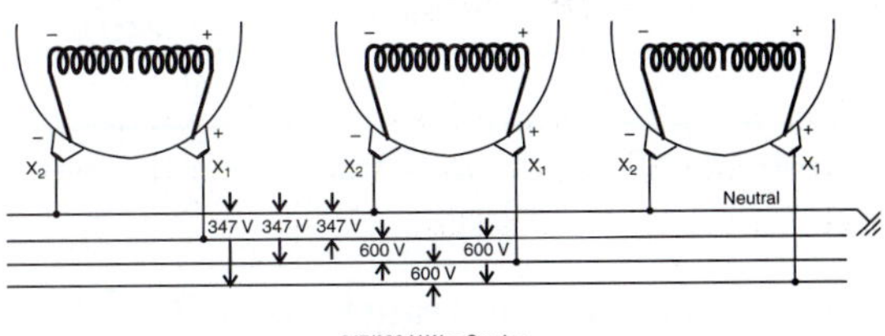 |

Figure 6–14   A three-phase 347/600V service.

6.4 Less Common Transformer Connections

6.4.1 Transformers Connected in Parallel

Two smaller transformers are sometimes connected in parallel to give the equivalent capacity of one large single-phase transformer. In Figure 6–15, the two secondary coils in each transformer are connected in parallel. One transformer feeds one leg at a positive polarity while the other transformer feeds the other leg at a negative polarity. Another option is to interconnect two transformers with the secondaries left connected in series. The positive terminals are connected to one leg of the bus, and the negative terminals are connected to the other.

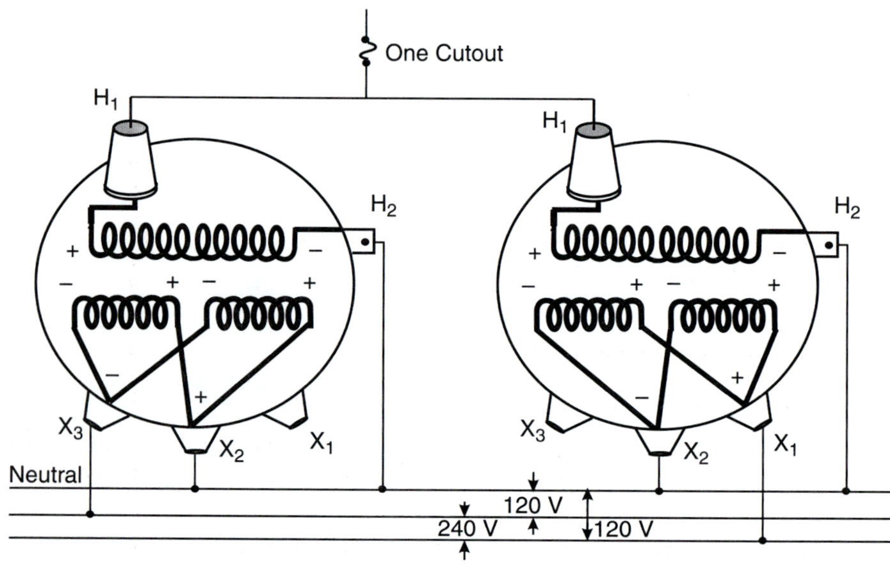

Figure 6–15 Single-phase transformers connected in parallel.

6.4.2 Open-Delta Transformer Banks

A three-phase delta service can be supplied with two single-phase transformers. Three primary wires are needed. A wye primary would need two phases and a neutral, and a delta primary would need three phases. This hookup is sometimes used as an economical way to feed a small three-phase delta load. One of the two transformers is often used as a *lighting transformer* and is sized to feed the single-phase portion of the load. The smaller transformer is often called the *power transformer* and is there to help provide the three-phase load, typically a motor. A three-phase wye-secondary service cannot be fed from two transformers.

When one transformer of a normal three-phase delta–delta or wye–delta transformer bank is found to be defective, the connections can be changed to the configurations shown in Figure 6–16, which restores the service as an open delta. The customer should be told to reduce demand on service until the transformer is replaced because the two good transformers will now have the capacity to supply only 57.7 percent of the capacity of three transformers.

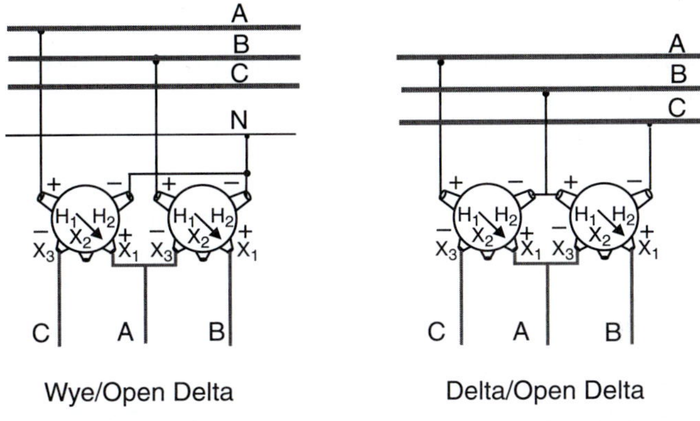

Figure 6–16 Open-delta transformer connections.

6.5 Secondary Networks

6.5.1 Single Phase Network

Most utilities feed their secondary bus radially from one transformer. The secondary bus will have bus breaks installed to prevent any interconnection with other transformers.

In a network system, a secondary bus can also be fed from many transformers connected in parallel to the same secondary bus.

- *When the primary of the transformer is opened, the primary terminal will remain alive because of backfeed from the live secondary. Open the network protectors or remove secondary leads to isolate a transformer completely.*

6.5.2 Three-Phase Network

A more common application, especially in a city core, is an underground network system.

- The load is divided among all the transformers connected to the bus.

- Fuses on the secondary bus (network protectors) between transformers isolate faulted transformers and interrupt only the customers near the defective transformer.

- All of the transformers on a common bus are connected with the same polarity.

- Because of the multiple transformers feeding into the same bus, a secondary fault is extremely explosive.

Three-phase transformer banks networked together to a common secondary must be similar:

- Each bank must have a similar impedance.

- Each bank must be on the same voltage tap setting.

- Each bank must have the same *angular displacement or phase shift*.

6.5.3 Angular Displacement Between the Primary and the Secondary

If	Then
Connecting a three-phase wye–wye transformer bank into a secondary network.	A wye secondary of a wye–wye bank, which has a 0-degree angular displacement, cannot be connected in parallel with a wye secondary of a delta–wye bank, which has a 330-degree angular displacement (see figure 6–17).

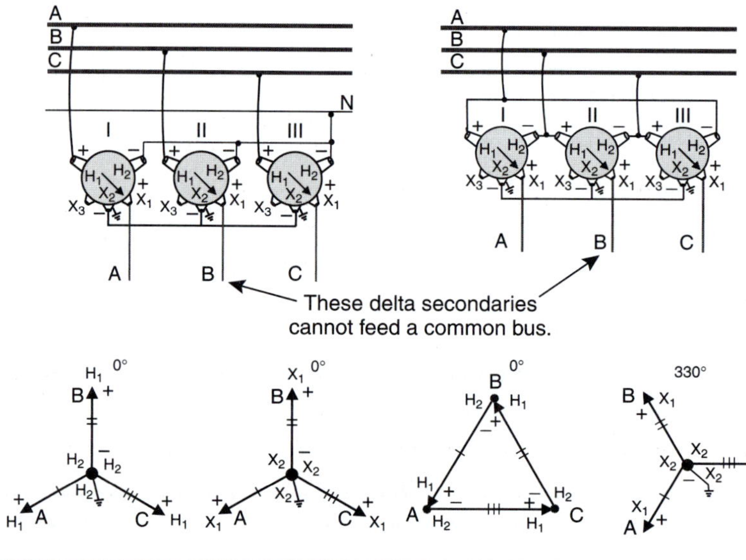

Figure 6–17 Angular displacement of a wye secondary.

(continued)

6.5.3 Angular Displacement Between the Primary and the Secondary

(continued)

If	Then
Connecting a three-phase delta–delta transformer bank into a secondary network.	A delta secondary of a delta–delta bank, which has a 0-degree angular displacement, cannot be connected in parallel with a delta secondary of a wye–delta bank, which has a 30-degree angular displacement (Figure 6–18).

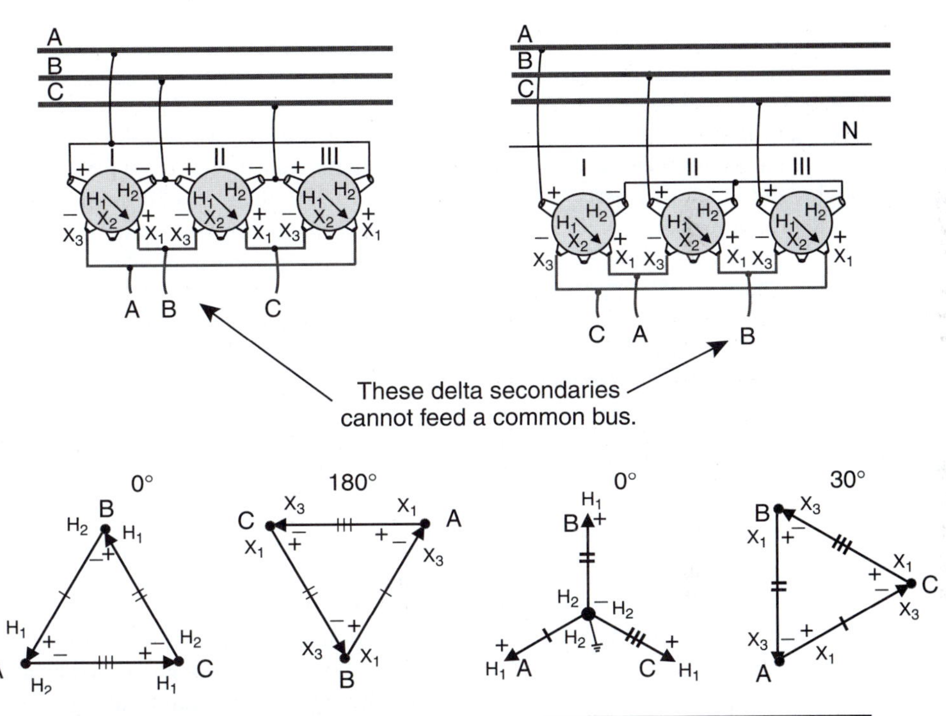

Figure 6–18   Angular displacement of a delta secondary.

6.6 Trouble Investigation: Transformers

6.6.1 Trouble Investigation: Secondary Voltage Irregularities

Trouble investigation for transformers mainly involves checking for problems on secondary services. Many *no-power* or *partial-power* calls are due to internal customer problems. A utility responsibility normally ends at the service entrance. Opening the main switch and taking a voltage reading will determine whether the utility is the cause of the problem.

Troubleshooting a Single-Phase, 120/240-Volt Service

If	Then
A customer has intermittent power and flickering lights.	Check for a voltage imbalance in the two 120-volt legs. If one 120-volt leg is two or more volts different from the other 120-volt leg, turn on a large 120-volt load. If the voltage increases on one leg and decreases on the other, there is a poor neutral connection. A loose connection on either leg can increase resistance to current flow and result in arcing and intermittent power.
The lights in part of a customer's premises are very bright, and in another part the lights are very dim.	A poor or open neutral connection blocks the normal return path from 120-volt appliances. The current travels back to the source through the other 120-volt leg or through 240-volt equipment. The 240-volt equipment operates normally.
A customer is receiving half power. Some of the lights work and some do not. None of the 240-volt appliances work.	One of the main fuses at the service entrance is likely blown. A voltage reading of about 120 volts between the top and the bottom of the fuse indicates a voltage difference and, therefore, that the fuse is blown. If the power is off, a continuity tester or ohmmeter also can be used to check the fuse. If the fuses are good, check the connections back to the transformer.
A customer complains about erratic or low voltage.	A poor connection could cause low voltage at various times. Heavy-duty equipment used by a neighbor on the same bus can cause voltage problems for others on the bus.

Troubleshooting a Wye Secondary (120/208-, 240/416-, or 347/600-Volt) Service

If	Then
A customer has an abnormal voltage.	Open the customer's switch and check the voltage. If the voltage on a phase to neutral reads 0, then check for a defective transformer or an open phase on the primary feeder.
	If the voltages on all three phases are balanced, close the customer switch and check the voltage. An unbalanced voltage indicates the customer's load is unbalanced.
Three-phase motors are overheating, and the thermal overload protection trips out the motor.	The usual cause of trouble on one phase is an unbalanced load. This type of service has single-phase loads. Unbalance beyond 3% can cause the low-voltage leg to draw more current and create heating of equipment.
	A faulty winding in a three-phase motor can cause a high current on a phase.
The load is balanced at the service entrance, but there is still a problem.	Open the customer's main switch and verify that the utility-supply voltage is correct.
	Close the customer's circuit breakers one circuit at a time and take voltage and current readings. A suspect circuit will have an unbalanced voltage or abnormal current readings for the equipment being fed.
All three-phase equipment and some single-phase equipment will not operate.	One phase is probably out. A phase-out can be due to a feeder problem, a transformer problem, or a blown fuse at the service entrance.
The single-phase equipment has intermittent power. The three-phase load is operating normally.	A poor or broken neutral connection will cause problems with single-phase loads.
	A customer-owned secondary (dry) transformer that steps down the higher 600- or 416-volt supply to 120/208 can cause problems.

Troubleshooting a Delta Secondary (600-, 480-, or 240-Volt) Service

If	Then
The customer has abnormal voltage. With the customer's switch open, the voltage on the supply side shows one or more of the phase-to-phase voltages at 0 volts.	One or possibly more phases feeding the customer is out of service. Check the transformers and the primary feeder. The customer's switch must be open for this voltage check because the backfeed in a delta service can show a voltage at the service entrance.
At the main panel, the phase-to-phase voltage readings are lower than normal and one phase-to-phase voltage reading is 0 volts. For example, on a 240-volt service, the readings are 210 volts, 210 volts, and 0 volts.	When one phase of a primary feeder loses power, the voltage at a wye–delta transformer bank will have a reduced voltage on two of the phase-to-phase readings and 0 volts on the third phase-to-phase reading. On a wye–delta transformer bank, the primary neutral is not connected to the system neutral or the secondary neutral, but is left ungrounded or floating. If the transformer neutral was grounded and one primary phase is opened, the transformer bank would become an open-delta bank and continue to supply three-phase power at a 57% reduced capacity, exposing the bank to a burnout because of an overload.
At the main panel, all three phase-to-phase voltage readings are normal, but one phase to ground is reading very low or 0 voltage.	A ground fault exists on a phase. When one phase of a delta circuit is faulted to ground, it does not blow a fuse because there is no path for a return current to the source. Both the phase and the ground are alive; therefore, a faulted phase-to-ground voltage reading will show a very low- or 0-voltage difference. Both the faulted phase and earth are alive in relation to the other two phases. A customer with a delta service often has a ground-fault indicating light connected between each phase and ground. The light will go out when a ground fault is present and the ground becomes alive.

6.6.2 Voltage Imbalance on a Three-Phase Service

A three-phase service supplied to a customer should not have a voltage imbalance exceeding 1 percent.

$$\% \ Voltage \ unbalance = \frac{Maximum \ Voltage \ Deviation \ from \ Average \ Voltage}{Average \ Voltage} \times 100$$

Where *average V* = average of the three voltages

Maximum Voltage Deviation from Average Voltage = voltage with greatest difference from the average

Example: Voltages readings are 220, 215, and 210. The average voltage is 215 volts. The maximum deviation from the average is 5.

The percent unbalance = 5/215 × 100 = 2.3 percent.

Any imbalance in a delta secondary will result in circulating currents within the service as the unbalanced current tries to find its way back to the source. To ensure that the utility is supplying a balanced voltage to the customer, the three transformers must have similar impedance and be on the same voltage tap.

An unbalanced secondary voltage can be caused by any of the following:

- An unbalanced customer load

- An unbalanced primary voltage

- Banked single-phase units with different kilovolt-ampere ratings

- Banked single-phase units with different voltage tap settings

- Banked single-phase units with different impedances

6.7 Specific Hazards Working with Transformers

Hazard	Controls
Reenergizing a defective overhead transformer can result in a violent expulsion of molten products from the fuse chamber or, in rare cases, a transformer explosion.	Test the transformer before reenergizing, or use precautions such as staying out from under the cutout, using a stick with an attached shield, and/or using an extra length of hot stick. A current-limiting fuse in series with the cutout fuse will reduce the risk of a violent transformer failure in locations where a high fault-current is available.
Bayonet-style fuses and current-limiting fuses in a dry-well canister have been known to fail explosively when used to energize a faulted transformer.	Energizing the transformer with a load-break elbow or from a remote location can reduce this risk.

(continued)

6.7 Specific Hazards Working with Transformers *(continued)*

Hazard	Controls
The magnitude of a fault on the secondary is very explosive when a short circuit is close to the transformer.	When work is required right at the secondary of a large transformer, consider isolating the transformer. Safety glasses should be worn when working on a live secondary because an eye can be permanently damaged by a large flash.
Ferroresonance can cause the voltage on a circuit to increase from two to nines times, causing equipment damage. The most common occurrence of ferroresonance involves a three-phase wye–delta or delta–delta transformer bank fed with a length of underground cable.	Ferroresonance can be prevented by using a three-phase gang-operated switch, or by keeping a load on the transformer when it is being energized or deenergized, or temporarily grounding the floating neutral of the wye primary, which will short out a possible series circuit.
On a wye–delta transformer bank, the H_2 bushings are interconnected but not connected to system neutral or grounded. The neutral connection is left floating (ungrounded) and, therefore, can have a high potential on it.	It must not be treated like a grounded neutral by anyone working on the transformer bank.

6.7.1 Protection from Transformer Backfeed

A secondary voltage feeding back into an isolated transformer is stepped up to a full primary voltage at the transformer primary terminal.

Sources of secondary backfeed are not always obvious and include the following:

- A portable generator connected into the customer system without opening the customer main switch

- A nearby recreational vehicle (RV) with a generator connected into home wiring

- An extension cord from a neighboring house

- When working on a transformer networked with other transformers to a common secondary bus, a live primary terminal (if no network protectors are opened) after the transformer is disconnected from the primary

Specific actions can be taken to provide protection from transformer backfeed:

- When working on a transformer, remove the secondary leads to ensure no backfeed is possible into the transformer's secondary terminals.

- Shorting out the secondary with approved grounds can be effective. Placing the grounds by hand can violate the minimum approach distance to the primary bushing.

- Removing meters from all customers fed from the transformer will eliminate backfeed from customer operations but not from other transformers connected to the same bus.

- A small (< 5 kW) portable generator feeding back into a grounded circuit will keep running and not short out. The resistance of the circuit between the generator and the grounds is too high to short out the generator. When working on a primary circuit, apply protective grounds. It can still be safe to work on the grounded conductors because the voltage at the grounded location has been lowered to an acceptable level. However, current will be flowing in the conductor, and the conductor must be jumpered before cutting or opening the circuit.

6.7.2 Neutral Connections Are not Ground Connections

The neutral and ground connections on a transformer may appear interchangeable because the two are interconnected. A neutral is a *grounded* conductor. Neutral connections are part of the electrical circuit and during normal operation carry the current back to the source. A ground wire is a *grounding* conductor. Ground-wire connections provide a path for current under abnormal conditions, such as during a lightning storm when an arrester or an insulator sparks over. The ground connections also bond the transformer tank and related equipment to keep them all at the same potential.

CHAPTER 7

Trouble Investigation: Outdoor Lighting

7.1 Locating the Cause of a No-Light Problem

Problem	Possible Causes
1. Lamp replacement is very frequent in some locations.	Incandescent lamps can fail prematurely if exposed to vibration or voltage surges. The capacitors may be shorted out. If so, replace the capacitors.
	A ballast with a higher wattage than the lamp will cause premature burnout of gaseous-discharge lamps. The nameplate on the ballast will show which type of lamp is matched to the ballast. A defective or incorrect ballast may be the cause, so be sure that a multitapped ballast is connected to the correct voltage.

(continued)

7.1 Locating the Cause of a No-Light Problem *(continued)*

Problem	Possible Causes
	Typically, incandescent lamps and series incandescent lamps should be replaced as a group yearly. Mercury-vapor, high-pressure sodium, and fluorescent lamps should be replaced as a group every 5 years.
2. The light stays on 24 hours a day.	Change out any photoelectric controller that is not fused because it is designed to stay closed when it becomes defective.
3. Lamp does not start.	*Lamp:* The lamp may be at the end of its life. Replace it. There could be a poor socket contact, so check for lamp tightness. *Photocell:* Check for a defective photocell by installing a replacement. *Low Voltage:* A low-voltage supply could be the cause, so check the voltage. *Ballast:* Make sure the multitapped ballast is connected to the correct voltage. Try a new ballast. For cold temperatures, install a ballast rated for a low-temperature start.
4. The lamp is on for awhile, goes out, and comes on again in a continuing cycle.	The lamp is near the end of its life and must be replaced. There could be a ballast-to-lamp mismatch. For example, a 175W lamp might be supplied by a 250W ballast. A low-voltage supply is unable to keep the arc in the lamp going. Check the supply voltage. Headlights from traffic are causing the photoelectric cell to turn off the light. Turn the cell to another direction. Make sure the proper lamp has been installed.
5. The light is out, even though it is good and the source circuit is alive.	Change out the probably defective ballast or test the ballast.
6. The lamp is flickering continuously.	The lamp is probably approaching the end of its life. Replace the lamp.

(continued)

7.1 Locating the Cause of a No-Light Problem *(continued)*

Problem	Possible Causes
7. The ballast is very hot.	The ballast temperature is higher under no-load conditions than under a loaded condition. The lamp is probably burned out. A thermal protector on the ballast should protect the ballast from being damaged.
8. The light does not appear as bright as it should.	Regular maintenance during group replacement of lamps should include cleaning the globe, as well as polishing the reflector and refractor. Check the output voltage of the ballast or replace the lamp.

7.1.1 Efficiency and Lamp Life of Various Lamps

Table 7–1 can be used as a guide to determine if lamps are failing prematurely.

TABLE 7–1   Efficiency and Lamp Life of Various Lamps

Type of Lamp	Energy Use	Lumens per Watt	Lamp Life Hours
Incandescent	High	8 to 18	Up to 2,000
Fluorescent	Medium	55 to 79	Up to 20,000
Metal–halide	Medium	38 to 75	Up to 20,000
High-pressure sodium	Low	72 to 115	Up to 24,000
Low-pressure sodium	Lowest	100 to 183	Up to 16,000

7.2 Resolving Some Specific Lighting Problems

Step	If	Then
1	There is no light.	Replace the lamp. Check for an arc-damaged socket or center contact.
2	There is still no light.	Check for power at the socket.
3	There is no power at the socket.	Check for power at the source leads of the ballast.

(continued)

7.2 Resolving Some Specific Lighting Problems *(continued)*

Step	If	Then
4	There is power at the source leads of the ballast.	If only one lamp is out, the lamp is not defective and the source is alive, so the ballast is most likely defective. An odor of burned insulation may be coming from the ballast. Test or replace the ballast. Ballasts cannot be repaired in the field.
5	There is *no* power at the source leads of the ballast.	Check the luminaire fuse (if equipped) and check the relay fuse or breaker in the source circuit.
6	The luminaire fuse (if equipped) is open.	Check for a short in the luminaire. Remove the lamp and do an insulation test at the source leads to the light using a 500 V insulation tester (megger or ohmmeter).
7	The source-side fuse of the relay feeding the streetlight bus is found open.	The relay is probably defective and must be replaced.
8	The load-side fuse of the relay is found open.	There is a problem downstream, on the streetlight bus or in a luminaire. Patrol the bus and/or do an insulation test at the luminaires. On a pilot-wire system, only the lights controlled by that relay will be out. On a cascading system, all the lights downstream will be out.
9	No faults have been found.	Check the photoelectric controller. Typically, defective and unfused photo controllers should keep the streetlight bus energized 24 hours a day. If the fuse on the source side of a fused controller is open, change out the controller. If the fuse on the load side is blown, patrol the streetlight bus and/or check for defective relays on the load side.

7.3 Testing a Ballast

Test	Details
Test the insulation of the ballast for short circuits.	To test for a short-to-ground, connect together all the primary and secondary leads and, using a 500V insulation tester (megger or ohmmeter), test between the leads and ballast case or ground. The insulation reading should be better than 0.5 megohms.
Test for proper output voltage using an open-circuit voltage test.	To check for proper voltage output, disconnect the secondary leads or remove the lamp to provide an open circuit. With the primary connected, the secondary voltage should be as stated on the ballast nameplate. Gaseous discharge lamps need a high voltage to start the initial electric arc. A ballast will step up a 120V supply voltage to provide a 750V to 4,000V output. For example, a high-pressure sodium lamp needs a high starting voltage of about 4,000V. After the lamp starts, the operating voltage of the lamp will vary depending on the type of lamp and wattage rating. Wait until the lamp is at normal brilliance, or approximately 30 seconds, before doing a voltage check on a gaseous-discharge luminaire. The voltage should settle to its normal operating voltage when the light is at normal brilliance.
Determine whether the ballast can limit the current as it is designed to do using a short-circuit test.	Remove the lamp and connect together the secondary leads to form a dead short. Apply a normal supply voltage to the primary and take a current reading on the shorted secondary leads. The purpose of a ballast is to limit the current, so very little current should be found between the secondary leads. Typically, the current for a mercury lamp should be 2A to 2.7A for a 175W lamp and 4.2A to 5.3A for a 400W lamp.
Check the primary coil for shorted turns or damage with a no-load current test.	Remove the lamp or open the secondary leads so that no load is on the secondary of the ballast. Apply a normal supply voltage to the primary and take a current reading on the primary leads. A reading of more than 50% of the rated primary current indicates a damaged ballast.

7.4 Checking the Power Source

7.4.1 Street-Lighting Controls

Check whether the streetlights are turned on and off individually or in groups using a control switch and a control circuit. They can be any of the following:

1. Individual lamp controls

2. Control switch with a pilot wire

3. Cascading relay system

Streetlight control switches that turn lights on and off can be photoelectric, a time-clock control, or manually switched. Photoelectric controls have taken over from manual switching or, in most cases, time-clock controls.

7.4.2 Individual Lamp Controls

Many streetlights and all yard lights are independently fed from any nearby and available 120/240-volt source controlled by an individual in-head photocell. If the light is not working, the source of the trouble will be at that light or power supply.

7.4.3 Control Switch with a Pilot Wire

A pilot-wire streetlight system has a separate wire (pilot wire) strung to all the streetlights to control many lights from one spot so that they all go on and off at the same time. Each individual light will probably not be supplied from the same source. It can be fed from any available and nearby 120/240-volt source. The pilot wire is strictly a wire to turn the lights on and off. It is energized from a 120/240-volt supply through a photocell and will activate a relay when it gets dark.

When the pilot wire is energized, the normally open relay contact connected to the pilot wire closes. Each relay in the system energizes a number of lights. A relay is just a switch activated remotely by the pilot wire. It is also possible to have a photoelectric control deenergize the pilot wire when it becomes dark. This releases the contacts of the normally closed relays, and the lights are energized. The normally closed relay system allows the lights to become day burners when the control switch or pilot wire becomes defective. Figure 7–1 shows normally open relays that are closed when the pilot wire is energized.

Normally open relay systems

1. One control switch, photoelectric or timed, energizes the pilot wire when it becomes dark.

2. The normally open relay contacts close when the pilot wire energizes the relay coil.

3. Each relay energizes a number of lights.

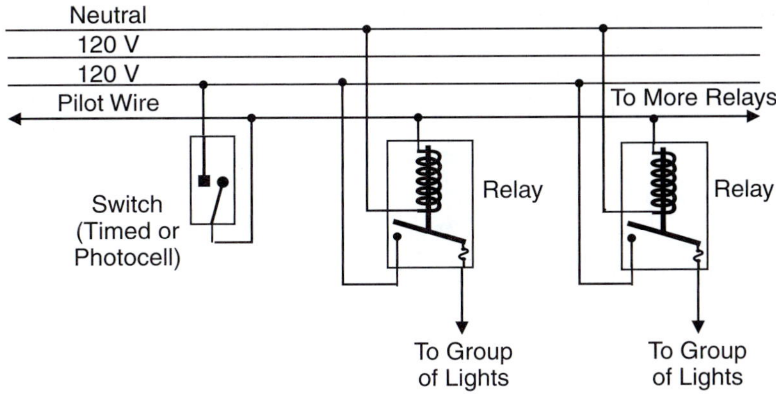

Figure 7–1   A pilot-wire control system.

Normally closed relay systems

1. One control switch, photoelectric or timed, deenergizes the pilot wire when it becomes dark.

2. When the pilot wire deenergizes the coil in the normally closed relay, the armature is released and the contacts close.

3. Each relay energizes a number of lights.

The normally closed relay system allows the lights to come on when the control switch or pilot wire becomes defective.

7.4.4 Cascading Lighting-Control System

A cascading relay system energizes and controls all the streetlights from one transformer and one control. This works well in areas with few or no existing transformers or secondary bus and works well on underground supply to lights. A streetlight wire is strung with relays installed along its length. To prevent having a very large relay (switch) to energize all the lights, each section of the lighting circuit activates the relay that energizes the next lighting circuit.

Figure 7–2 shows that when a relay is energized it will energize a pilot wire to energize the next relay. The length of the feed will be limited by the voltage drop that will occur when the feed becomes too long for the size of wire.

7.4.5 Series Outdoor Lighting

In a series circuit, lights are fed with one relatively small and continuous wire that starts at a source transformer, forms a loop through the streets being illuminated, and goes back to the transformer. The series circuit wire is fed at a primary voltage and is insulated and placed in a primary position on overhead distribution poles. For an underground installation, the cable would be

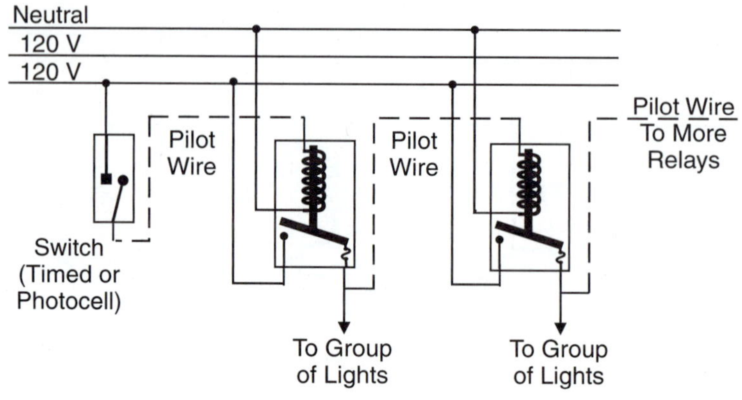

Figure 7–2   A cascading control system.

a primary voltage cable. A high voltage is needed because in a series circuit the sum of the voltage drop across each individual light is equal to the source voltage. A special constant current transformer, also called a "regulating transformer," is used to raise and lower the voltage automatically, depending on the number of streetlights in the circuit. A constant current has to be maintained because if one lamp burns out the total resistance of the lighting circuit becomes smaller and the current tends to increase. The voltage will go as high as needed (up to several thousand volts) to supply a constant current through the circuit.

As light units are added, the primary voltage will go up to continue to provide a constant current (typically 6.6, 7.5, or 20 amperes) through the special series circuit lamps.

A hazard to powerline workers is that the voltage across any break in a series circuit will be equal to the output voltage of the transformer. Maintenance should be done with the circuit isolated and grounded to ensure that a circuit is not opened, cut, or joined unless a break is "jumpered" to avoid an open-circuit condition. Because a primary voltage would be present across any open point in the circuit, the lamp is designed to automatically bypass the current. Figure 7–3 shows the special series circuit lamp with the thin film of insulation between two points on the lamp that is punctured by the high voltage restoring the lighting circuit across the break. A lamp used in a series circuit has a heavier filament because it has to carry a relatively high current compared to lamps connected in parallel. The socket in most series light fixtures is designed so that removing a bulb will not open the circuit.

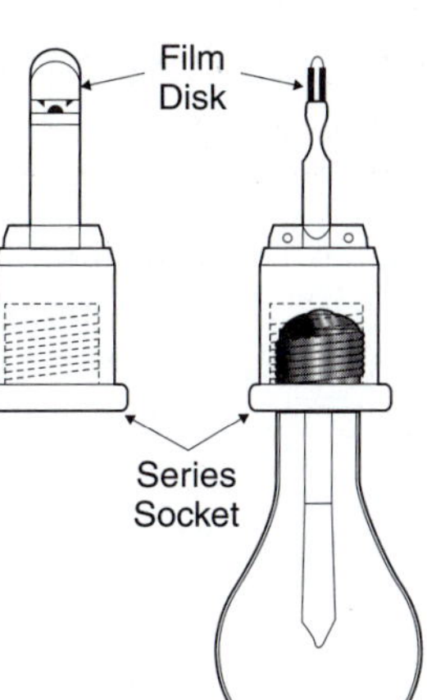

Figure 7–3   Special series circuit lamp.

7.5 Safety and Environmental Hazards

Hazard	Control
Ultraviolet-radiation hazard	The inner tubes of mercury-vapor and high-pressure sodium vapor lamps put out an intense ultraviolet radiation. The outer phosphor-coated bulb protects people from exposure to the ultraviolet radiation. If exposed, the eyes will develop the same "sand-in-the-eye" symptoms as they do from exposure to an electric arc or welding arc. Even though the outer bulb is broken, it is possible for the inner bulb to continue to emit an unseen ultraviolet radiation without the outer bulb acting as a filter. The outer bulb can break while replacing a lamp and expose the worker to ultraviolet burns from the inner bulb. Disconnecting the power source will protect a person from ultraviolet-radiation burns.

(continued)

7.5 Safety and Environmental Hazards *(continued)*

Hazard	Control
High-voltage hazard	Gaseous-discharge lamps need a high voltage to start the initial electric arc. A ballast will step up the 120-volt supply voltage to provide a 750- to 4,000-volt output. For example, a high-pressure sodium-vapor lamp needs a high starting voltage of about 4,000 volts. After the lamp starts, the operating voltage of the lamp will vary depending on the type of lamp and its wattage rating. Wait until the lamp is at normal brilliance (or approximately 30 seconds) before doing a voltage check on a gaseous-discharge luminaire. The voltage should settle to its normal operating voltage when the light is at normal brilliance. Some typical operating voltages are found in Table 7–2.
High-voltage hazard working with series lighting circuits	Series lighting circuits are not secondary circuits. The source-constant current transformer can supply several thousand volts, depending on the number of lights in the circuit. The voltage across a break in the circuit will equal the full-line voltage. If the series loop is to be opened or repaired, the risk of electrical shock is reduced by opening the source transformer and grounding at the point of work.
Toxic-chemical hazard	Mercury, sodium, and phosphor are toxic substances. These chemicals can be found inside the outer bulbs of gaseous-discharge lamps and are released when the outer bulbs are broken. Staying outdoors and upwind when working with these lights reduces the risk of exposure. *Mercury* is hazardous to the human body. It is usually in a gaseous state because it vaporizes at 14°F (10°C). In the gaseous state, it can easily enter the body through the lungs. *Sodium* is a highly active poisonous chemical that oxidizes very quickly in air. Sodium in contact with water produces a chemical reaction that creates sodium hydroxide (caustic soda), which can damage the lungs when inhaled. *Phosphors* are poisonous and must not be inhaled.

(continued)

7.5 Safety and Environmental Hazards *(continued)*

Hazard	Control
Environmental hazards	The capacitors used in older luminaires (before 1979) are impregnated with polychlorinated biphenyls (PCBs). PCBs are not biodegradable and will stay in the environment and end up in the food chain. PCB waste must be disposed of properly according to regulatory requirements. The mercury and lead found in lamps are toxic substances. Some companies will separate and recycle lamp components, such as glass, metal, mercury, and phosphor powder.

TABLE 7–2   **Typical Operating Voltage for Gaseous-Discharge Lamps**

Lamp	125W	150W	250W	400W	700W	1,000W
Metal–halide			100V	120V		250V
Mercury–vapor	125V		130V	135V	140V	145V
High-pressure sodium-vapor		100V	100V	105V		110V

CHAPTER 8

Working with Underground Systems

8.1 Specific Hazards Working with Underground Cable

Hazard	Controls
Breaking or opening sheath continuity	In a looped circuit, any break in the sheath of a cable can have a high voltage across the break. When work is performed on or near cable, the metallic sheath (shield), continuity has to be maintained. The sheath should be treated the same way a neutral is treated on overhead systems.

(continued)

8.1 Specific Hazards *(continued)*

Hazard	Controls
Distinguishing between the cable to be worked on and other cables buried in the same trench	Even with every reasonable assurance that the proper cable has been identified, the cable should be "spiked" before it is worked on. A spiked cable will cause a short circuit if the cable is hot, and it provides a positive physical identification of a desired cable.
Accidentally energizing a faulted cable	When closing in on a faulted cable, the high fault current generated can become a ground-gradient hazard near the fault, can develop into an explosive hazard in a confined space, and can damage an entire length of cable beyond repair. Very few transient faults occur on an underground cable. A blown fuse generally indicates a cable fault and that a test for the existence of a fault should be administered before reenergizing.
A cable with the sheath grounded at one end can cause a potentially high voltage between the sheath and the neutral during a fault condition at the other end.	Engineering specifications should include a requirement to flag (warning signs) this hazard. See section 8.1.1 for a more detailed explanation.

8.1.1 Unacceptable Electric Current or Voltage on Sheath

The electromagnetic field in an underground cable can induce unacceptable levels of current and voltage on the metal sheath and concentric neutral wires. The magnetic field from the current-carrying conductor induces a current in the metal sheath or concentric neutral wires when the sheath is grounded at both ends of the cable. Figure 8–1 shows that when the sheath is grounded at both ends, a circuit is created for the induced current. The induced current in the sheath flows back to the source through earth. The current on the sheath can be high enough to cause heating and derate the ampacity of the cable.

Planning engineers sometimes specify grounding a metal sheath at only one end to reduce heating caused by induced current. When only one end of a sheath is grounded, the circuit is open and no current can flow in the sheath. An additional conductor must be strung to serve as a neutral for the circuit.

At the end where the sheath is not grounded, a potentially high voltage runs between the sheath and the neutral (see Figure 8–2). The voltage is generally less than 40 volts. Under fault conditions, the voltage across the open point be-

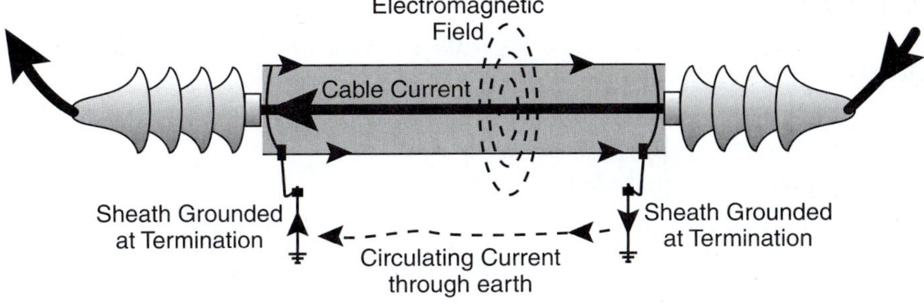

Figure 8–1   Circulating current in sheath.

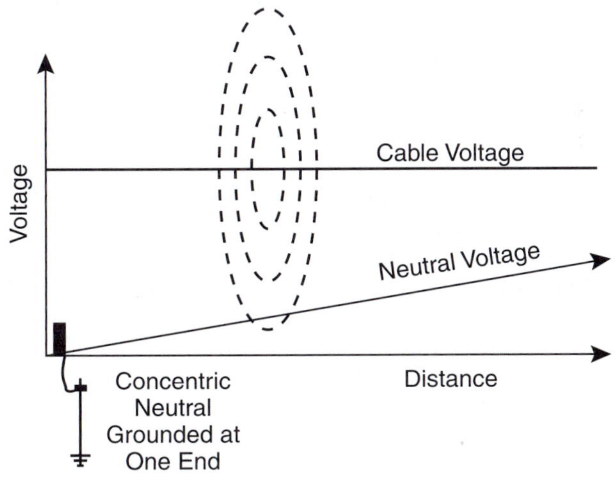

Figure 8–2   Voltage rise in concentric neutral.

tween the sheath and the neutral could be several thousand volts and hazardous to anyone working at the termination.

Specifications should include a requirement to flag the open circuit location as a hazard. A surge arrester could also be specified to limit the voltage across the open point.

8.2 Tests on Underground Cable

8.2.1 Locating and Tracing an Underground Cable

Cable locators consist of a transmitter, which induces an electromagnetic field (radio frequency or audio frequency) on the target conductor, and a receiver that traces the field. The methods presented in this section do not apply well to cables in duct banks under city streets. However, cables in duct banks tend to be more protected from faults, and better data are available on their routes and locations.

The receiver, a directional antenna, is swept from side to side. The strength of the electromagnetic field increases as the receiver gets closer to the cable. The electromagnetic field can be distorted by other cables, a turn, a change of depth in the target cable, or other metal, such as property stakes. Use the "Null" mode to locate the cable and the "Peak" mode to verify the location.

Safety precautions include the following:

- Ensure that the circuit to be tested is isolated. Reduce the risk of applying leads to a live circuit. Take out a formal clearance, test and ground, apply the test leads, and then remove the grounds.

- A cable is a capacitor, and it will have a voltage on it long after a high-voltage test is completed. Ground while making the connections for a test, and ground the cable before removing the test equipment. After a DC test is applied, it takes a long time to drain the charge off the cable. Wait 15 minutes, then ground it.

- Put up a barrier at both ends of a cable while it is being tested.

Three Methods of Putting an Active Signal on a Cable

Method	Steps	Additional Information
1. The transmitter is attached right onto the sheath or to an isolated core conductor with the *direct-connect method* (see Figure 8–3).	1. The transmitter can be connected to the sheath if the cable has an insulated cover over it. To trace a cable with a sheath that is continuously grounded (e.g., lead cable), use the lowest possible frequency. 2. Ground the far end of the cable to complete a circuit so that the signal current returns to the transmitter ground through earth. 3. Set to a low frequency (less than 10 kHz). A lower frequency is less likely to spill over (couple) to other buried utilities and the signal travels farther.	1. This is the most accurate method, because the signal (tone) is isolated to one cable. 2. Can be connected to the sheath of a live cable *but not to a live cable conductor.* 3. When attaching the transmitter to a sheath at a riser pole with many cables, the signal current will take the easiest path, which may not be the target cable.

(continued)

Method	Steps	Additional Information

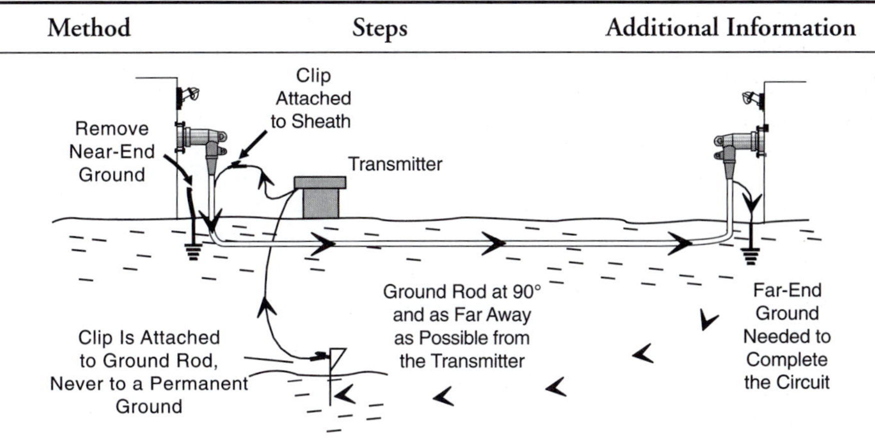

Figure 8–3   Tracing cable: direct-connect method.

2. The *general-induction method* induces a signal into a cable from a transmitter set on the ground directly over the cable (see Figure 8–4). As an operator with a receiver passes over the cable, the receiver picks up the tone so its route can be traced.	1. The sheath must be grounded at both ends. 2. The receiver should be at least 15 paces away from the transmitter to avoid receiving a signal directly through the air. 3. Try a medium frequency (30 kHz to 90 kHz) to reduce interference from other sources. A high frequency can be used to sweep a large area when looking for all buried utilities.	1. The transmitter can also induce a signal on other cables, pipes, etc., causing the receiver to pick up false signals. This is not a good method for congested areas. 2. For this method, the cable sheath has to have an insulated cover over it to prevent the signal current from following other ground paths. 3. This method is not suitable for cable deeper than 6 ft. (2 m).

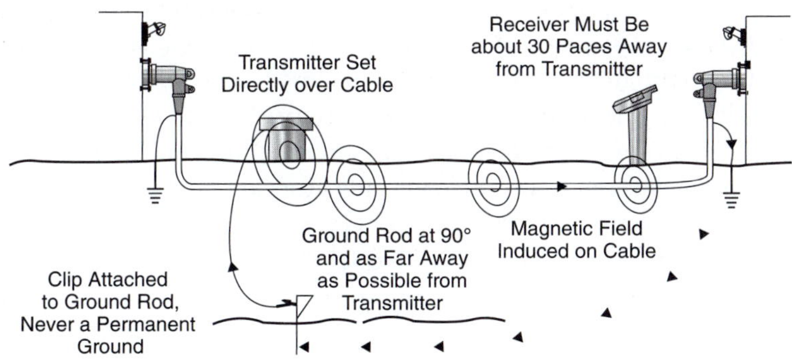

Figure 8–4   Tracing cable: General-induction method.

(continued)

Method	Steps	Additional Information
3. The *inductive-coupling method* uses a C-clamp (coupler) to concentrate the magnetic field signal onto the sheath or conductor (see Figure 8–5).	1. To use the inductive-coupling method, the cable sheath must be grounded at both ends to form a complete circuit for the signal current. For secondary cable without a sheath, a ground is placed on each end. 2. Use radio frequencies. Audio frequencies will not work.	1. The C-clamp can be connected over the sheath of live cables or over live secondary cables. 2. The concentration of the signal on a cable ensures less spillover to other cables on a riser pole or in a trench.

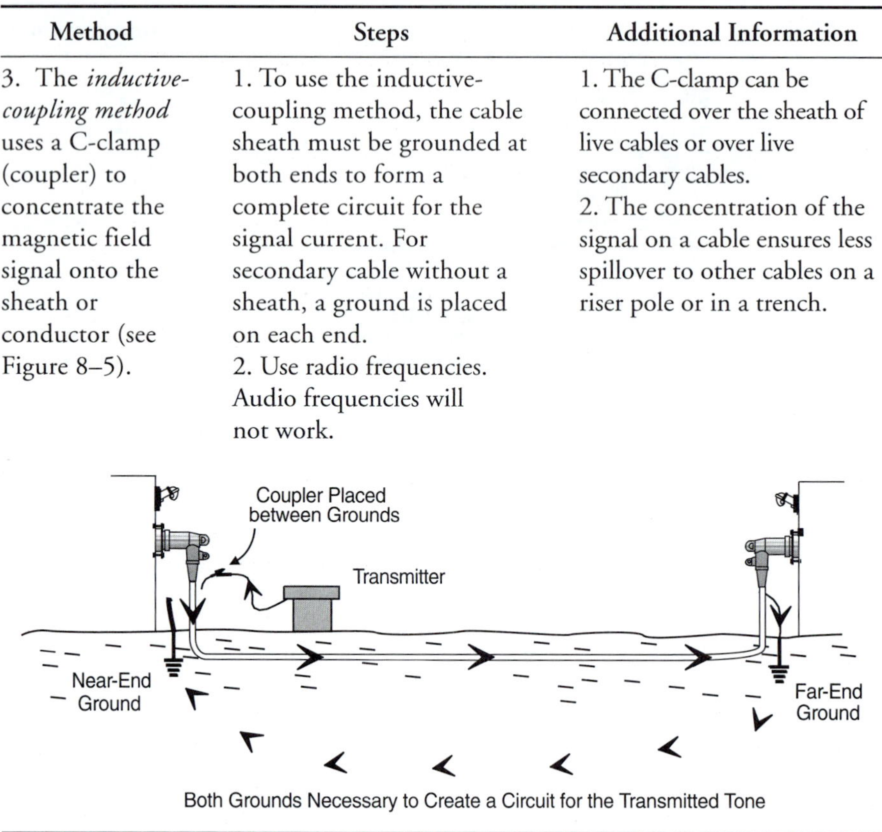

Figure 8–5   Tracing cable: Inductive-coupling method.

8.2.2 Locating a Faulted Section of Cable

When one section of cable has a fault, a fuse or breaker somewhere in the system can trip out many sections of cable. To find the faulted section of cable, a line crew has to progressively open, test, and close (sectionalize) each section of cable until it finds the fault, unless fault indicators have been installed.

A fault indicator will give a visual "flag" after the high magnetic field from a fault current triggers the unit. The fault indicators are put on the cable at various strategic locations (see Figure 8–6). The system can be traced from the source through each fault indicator until one is found in the "no-trip" position. The fault indicator in the no-trip position did not have a fault go through it, therefore, fault is upstream.

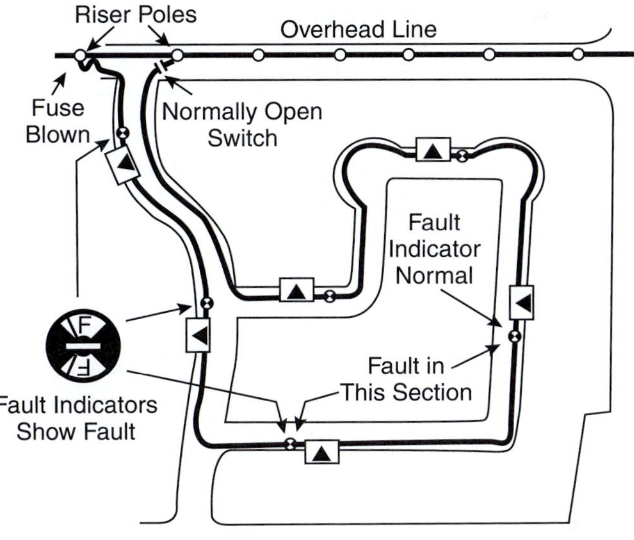

Figure 8–6   Fault indicators in an underground system.

8.2.3   Verifying the Existence of a Cable Fault

A cable fault will be caused either by some kind of insulation breakdown (anywhere from a dead short to a high-resistance fault) or a break in the conductor (open circuit). A megohmmeter (insulation megger, *not* an earth megger) can be used to test the insulation resistance of the cable and to verify if there is in fact a fault that cannot be pinpointed. This is a nondestructive test if the proper test voltage for the cable insulation is used—for example, use a 500-volt megger on a secondary cable rated up to 600 volts. Typical fixed DC voltages on a megohmmeter are 500, 1,000, 2,500, and 5,000 volts.

The faulted section must be grounded to drain any capacitance in the cable before any testing is done. Using rubber gloves, the concentric neutral can be isolated from ground at both ends to prepare for testing the cable. The insulation can be tested in the following ways:

1. Between the isolated neutral (sheath) and the grounded phase conductor

2. Phase to phase

3. Between the target conductor and other grounded conductors

The voltage is applied for about 1 minute until the charging effect has diminished and the reading is stabilized.

Typical Megohmmeter Readings

Table 8–1   Typical Megohmmeter Readings

Meter Reading	Type of Fault	Notes
0	A shunt fault, where the fault path is from the high-voltage conductor to ground or to the sheath (a short circuit or dead short).	Look for sources of potential dig-ins, etc.
More than 0 (hundreds of ohms resistance)	A low-resistance shunt fault allows current to flow to ground with very little resistance. Over-current protection will trip out the circuit.	The location of this type of fault is relatively easy to pinpoint with standard instrumentation.
Less than ∞ (thousands of ohms resistance)	A high-resistance shunt fault where often both the center conductor and neutral are still intact. (A cable is considered good if it reads 1 megohm for each kilovolt of cable rating, i.e., 5-megaohms 5-kilovolt cable.)	This is the most common fault because very fast circuit protection tripped the circuit before the cable was damaged more. Circuit protection may hold for a while when the cable is reenergized.
Infinity ∞	Indicates a series fault where the conductor and/or neutral are burned open, or it indicates a cable without a fault.	Ground at the far end of the cable and measure again. A "0" resistance would indicate that it is not an open circuit.

8.2.4 Pinpointing the Location of a Cable Fault

After a megohmmeter reveals the most likely type of fault in a cable, choose the instrument suitable for finding that type of fault. The methods described here are most suited to direct-bury cables and generally not suitable for cable faults in duct banks under city streets.

8.2.4.1 Short Circuit or Low-Resistance Fault

This is the easiest fault to find. The cable tracing equipment described in section 8.2.1 can be used. After all the cable grounds have been removed, a signal current from a transmitter is forced out where the cable insulation has failed. The voltage gradient above the fault is picked up by a probe or A-frame connected to the receiver. The A-frame is moved (Figure 8–7) and turned in the vicinity of the fault until the signal strength at each leg of the A-frame is equal (a "null" signal), indicating that the A-frame is directly over the fault.

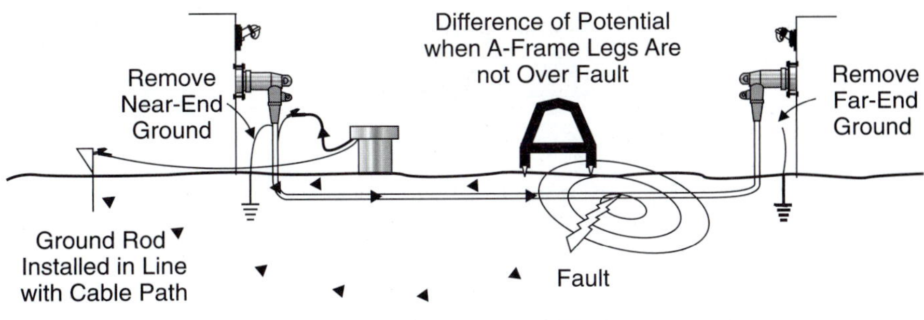

Figure 8–7 Pinpointing a cable fault.

Cable radar (TDR—Time–Domain Reflectometer) can be very effective as it sends a high-frequency pulse along the cable and measures the time it takes for the signal to be reflected back. Irregularities in the cable and their locations can be shown in graph form on a monitor screen. Short circuits, open circuits, splices, transformers, and other irregularities will show up on screen in waveforms that trained operators can recognize. After training, a person can determine the distance to the fault with cable radar.

A *thumper* (surge generator, impulse generator, capacitive discharge set, banger) finds the short circuit by storing a high-voltage charge in its capacitor and then discharging it into the cable to cause a flashover and an acoustic shock wave at the fault location. The shock wave can be detected by vibration, sound, seismic detectors, or microphones.

8.2.4.2 High-Resistance Fault

High-resistance faults (the most common type) are also the most difficult to find. They are often caused by a faulted conductor burning back toward the source and the cable core, surrounded by the insulation, is "healed over" and will therefore become a high resistant fault. It may be easier to find a high-resistance fault from the load end of the cable. A high-resistance fault can be broken down into a low-resistance fault by a thumper or burner.

Caution: An aged cable that has been thumped excessively is a likely candidate for a failure within the next year. The voltage chosen should be below the rated cable voltage and generated for as short a time as possible to reduce the risk of damaging the cable at other weak spots.

The least damaging and most effective approach to finding a high-resistance fault would be to connect a thumper, cable radar (TDR), and an arc reflection method (ARM) filter (see Figure 8–8). The thumper "turns on" the fault momentarily as the cable radar captures the image of the brief flashover at the fault location. The ARM filter is needed with this method to prevent the thumper output from damaging the cable radar instrument. Damage to the cable is limited because frequent thumping is not needed and the voltage will not be much

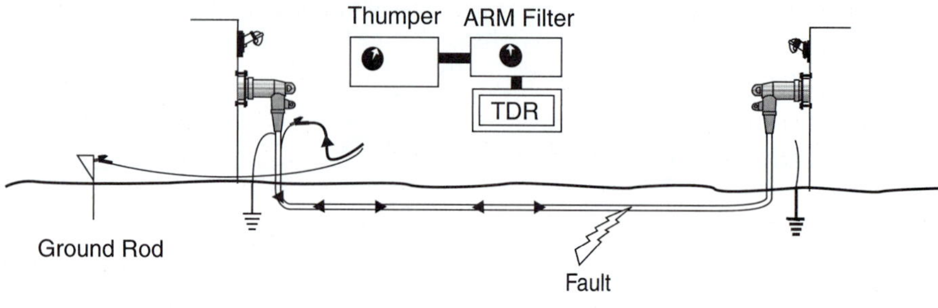

Figure 8–8   Arc reflection method.

beyond the breakdown level as the fault breaks down before the voltage rises too high.

8.2.4.3 Open Circuit

If an open circuit has no leakage to ground, TDR is effective for pinpointing the location of the open point. An open circuit leaves a distinct blip on the monitor screen (see Figure 8–8).

8.2.5 One Source of an Erratic Secondary Cable Fault

A fault on a secondary underground cable often starts out as a complaint of erratic power. A voltage reading at the customer may show a normal voltage. A break in the insulation of a secondary cable is often a high-resistance fault. When water gets into an aluminum conductor, it slowly corrodes the aluminum and turns it into an aluminum hydroxide, which is a high-resistance, white, powdery material. The voltage may be normal until a load is put on the conductor because the corroded higher-resistance cable will cause an increased voltage drop as the load increases.

Use a 500-volt insulation tester to test the service. A reading of anything less than 1 megohm is an indication of a faulty cable. The insulation tester has the ability to send a 250-volt, 500-volt, or 1,000-volt test voltage for insulation resistance testing.

8.2.6 Testing Cable

A new or repaired cable installation should be tested before it is put into service. One method is to apply an "over-voltage" (hipot) test. Shortening the life of the cable with this test method is of concern, and some utilities will not use this test.

A DC or the preferred very low frequency (VLF) hipot test of a cable measures any leakage current in milliamperes between the phase conductor and the sheath or earth. The level of leakage current is not critical if it remains at a steady level. If the leakage current increases rapidly during the test, it will indicate a potential fault. To avoid damage to the cable, stop the test if the leakage current begins to increase. Of course, you now will have a high-resistance fault to pinpoint.

A DC hipot adapter attached to a phasing tool can test a cable with an AC voltage when using an existing live AC source and converting it to DC. The DC voltage impressed on the cable will be at about the peak value of the source AC voltage. The hipot adapter is installed on the meter stick (see Figure 8–9). This stick is connected to a live source. The live source can be a transformer bushing or an open cutout at a riser pole. Some bushing adapters and elbow adapters allow making the connection. The reel stick is attached to the cable to be tested. The cable must be isolated at both ends because a DC signal through a transformer will show up as a short circuit.

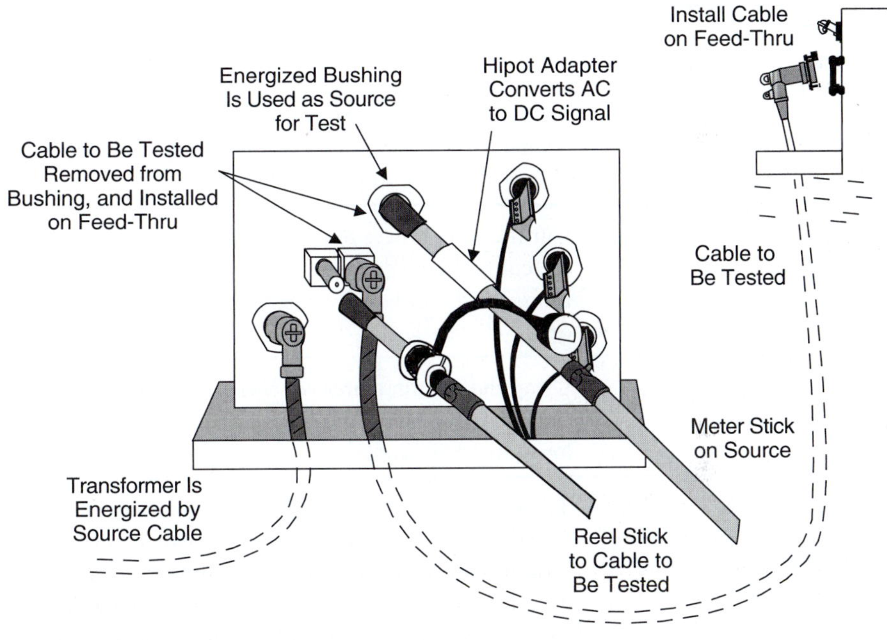

Figure 8–9   Phasing tool used as hipot tester.

8.3 Splices and Terminations

8.3.1 Splicing and Terminating Principles

Although there are a lot of cable types and a lot of different splicing and terminating kits and tools, the principles applied are the same for all. For a splice, this means exposing the inner core so that the core conductor(s) can be spliced and then rebuilding the cable so that a conductor shield, an insulation shield, a neutral, and a jacket are installed in such a way that no additional electrical stress will be placed on the insulation. For a termination, this means exposing the core conductor and making an electrical connection to some type of exposed terminal and then rebuilding the conductor so that the exposed end will have a long tracking distance back to the neutral and the semiconducting conductor shield in such a way that no additional electrical stress will be put on the insulation.

8.3.2 Procedure for Splicing a Single-Conductor Concentric Neutral Cable

The following procedure is generic and is intended to outline the fundamentals for making a splice or termination. Always follow the owner's, employer's, or manufacturer's specifications.

Step	Action	Details
1	Train the cable into position.	The cable must be trained into position in such a way that it curves with the natural curve in the cable. A sharp bend can leave voids between the insulation and the conductor shield and can leave electrical stress points at the bend.
2	Cut the cable.	Cut the cable where the splice is to be made. Because the core is being exposed, it is at this stage where the tools, hands, and environment must be kept clean and dry.
3	Lay out the dimensions.	Using the manufacturer's or owner's specifications, prepare to remove the jacket and so on at the exact location on the cable.
4	Remove the cable jacket or sheath.	There are many types of cables and many types of jackets. The surface of the cable jacket or sheath should be cleaned to remove all pulling lubricants, oils, dirt, and so on that can later contaminate the splice. Cleaning a lead sheath can mean using a shave hook or rasp until it is shiny. A polymer jacket can be cleaned by using an abrasive cloth. The finished splice will overlap onto the cleaned jacket. Use the special tools available for removing the jacket for the distance specified, with the understanding that the cable shield and the insulation must not be cut or damaged. Clean the removed jacket and sheath material from the work area.

(continued)

Step	Action	Details
5	Remove the metallic shielding.	The metallic shielding is often composed of wires that must be controlled. Tape hose clamps; other methods can be used. Depending on the location of the cable and other live cables, the metallic sheaths of the two cables being joined should be jumpered. Start each strand of the wire sheath at the end of the cable, usually with a knife, while being careful not to nick the cable insulation.
6	Remove the semi-conducting insulation shield.	The semiconducting material that makes up the conductor shield must be removed completely from the insulation. It is a good practice to leave 1/4 in. (50 mm) of the material where it meets the metallic shielding. It is easier to stop there than tight to the end, and it serves as a reminder not to overlap it later with insulating tape or other material. The semicon material must be scraped off to leave the insulation clean. If cleaning with solvents or other cleaners, ensure that the cleaner is compatible with the cable insulation. When cleaning, wipe away from the end and toward the semicon.
7	Remove and pencil insulation.	The insulation is removed from the conductor core for a distance of half the sleeve or connector length, plus an allowance for a short length of exposed conductor between the insulation and the sleeve. Some materials are very hard to cut, and the application of heat from a hair dryer will ease the task. The insulation at the sleeve end is penciled to a specified distance with a knife or, preferably, a penciling tool. The space between the sleeve and the penciled end of the conductor is used to make an easier transition for the shielding tape.
8	Splice or connect the conductor.	When using a sleeve to join the conductors, use the sleeve, dies, and press specified for the type of conductor and in the splicing kit specifications. When splices are made between two different sizes or between copper and aluminum, sleeves made intentionally for that purpose must be used.

(continued)

Step	Action	Details
		If using a soldered sleeve, follow the instructions to prevent overheating and to allow for the proper cleaning of the joint.
		Make the joint smooth. Be sure to remove any sharp edges on the joint.
9	Apply new conductor shield—semiconducting tape.	The conductor shield is conductive, prevents any electric stress from concentrating in one area, and fills any small indent at the sleeve or connector. Start the semiconducting tape at about 1/8 in. (3 mm) up the penciled slope, and tape a smooth layer back across the joint to the same position on the other slope.
10	Clean the exposed cable insulation.	Remove all dust and contamination with an abrasive cloth or an approved solvent.
11	Apply insulating tape.	Use calipers to measure the diameter of the sleeve/connector on the core, then double the width of the caliper. The doubled distance will be the thickness that the insulated tape will be applied over the connector. Apply tape to build up the insulation while stretching it to three-quarters width. Apply tape over the sleeve so that it is built up to the thickness set by the calipers while sloping the tape back to the original cables.
12	Apply the insulation shield—semiconducting tape.	Apply the semiconducting tape, while overlapping by half for the full length of the splice until it covers the existing semiconducting layer on each end of the splice by 1 in. (2.5 cm).
13	Apply a protective jacket.	On a taped splice, the protective jacket would be two layers of PVC tape, half overlapped over the full length of the splice.
14	Connect the two ends of the metallic concentric neutral.	The neutral is joined with a specified sleeve or connector. If a jumper had been installed earlier it can be removed after the joint is made.

CHAPTER 9

Installing Personal Protective Grounds

9.1 Three Reasons to Install Personal Protective Grounds

Purpose	Details
1. Install personal protective grounds to prove isolation.	After isolation and testing, the installation of grounds is the final proof that you are about to work on the correct circuit.

(continued)

9.1 Three Reasons to Install Personal Protective Grounds
(continued)

Purpose	Details
2. Install personal protective grounds to have protection from accidental reenergization.	When working on a grounded circuit, it is necessary to have protection from accidental reenergization, which can come from operator error, contact with neighboring circuits, lightning, or backfeed.
3. Install personal protective grounds to provide protection from induction.	Grounds provide protection from two kinds of induction hazards. *Electrostatic induction* will induce a voltage on an isolated circuit. *Electromagnetic induction* can induce a current flow in the conductor.

9.2 Personal Protective Grounds Control Current and Voltage

9.2.1 Apply the Grounding Principle to Control Current

The Grounding Principle: Personal protective grounds are installed to reduce any current flow through a worker to an acceptable level by providing a low-resistance parallel shunt around the worker. At the same time, if the circuit is or becomes energized, the grounds must be big enough to withstand any fault current in the circuit.

Control Current Flow	How
1. Use personal protective ground cable and clamps that are designed and large enough to carry the available fault without failing.	The cable and clamps must be able to withstand the fault current available at the location where grounds are installed. The fault current available on a circuit and the protective grounding cable and appropriate clamp sizes must be specified by utility engineering. Clamps used for circular-type conductors are not interchangeable with clamps used on flat surfaces. Table 9–1 is a sample of the grounding cable sizes needed to withstand a high fault current.
2. Provide a very low-resistance shunt (parallel path) around a worker so that all but a harmless amount of current passes through any parallel path established by a worker.	When a person on a structure is in contact with a circuit that is accidentally reenergized, the protective grounds and the person form two parallel paths to ground. To provide a good shunt around the worker, the protective grounding set must be kept in excellent condition by testing regularly.

(continued)

Control Current Flow	**How**
	Apply protective grounds in a T pattern (see Figure 9–1). The three-phase fault condition imposed by the grounds on the circuit in a T pattern will reduce current flow in the cable connected to the grounding electrode and will thereby reduce the proportion of current flow through a worker in parallel to ground.

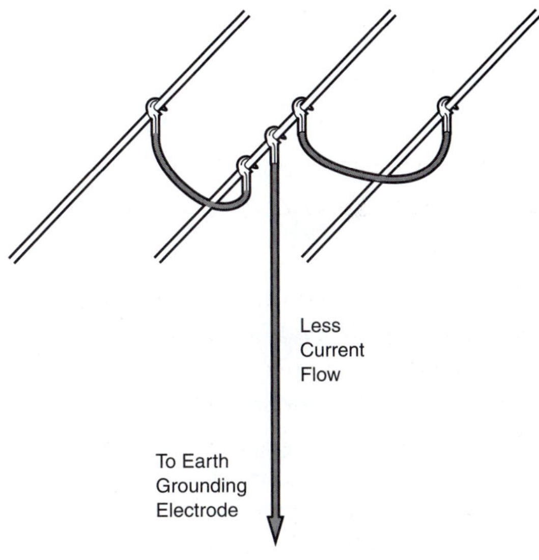

Figure 9–1   Protective ground installed in a T pattern.

As discussed in section 9.2.2, equipotential grounding will virtually eliminate current flow through a person that is a parallel path.

Control Current Flow	**How**
3. Promote a fast trip-out of circuit protection.	Connect the protective grounds to a good ground electrode. The best ground electrode is a permanent ground network such as a neutral, station ground, or tower steel.
	Circuit breakers, reclosers, and fuses open when a current feeding a fault goes through these devices for a specified period of time.
	In the case of reenergization, grounds on all three phases will provide an excellent phase-to-phase fault that will activate the protective switchgear.

(continued)

Control Current Flow	How
	If one phase is energized, the protective grounds must be connected to the best ground electrode available.
4. Install one set of personal protective grounds at the point of work or if more than one set of grounds are needed on the circuit then install them in an exact sequence.	Circulating currents are set up when a second set of grounds is installed and creates a circuit for the induced current, as shown in Figure 9–2.

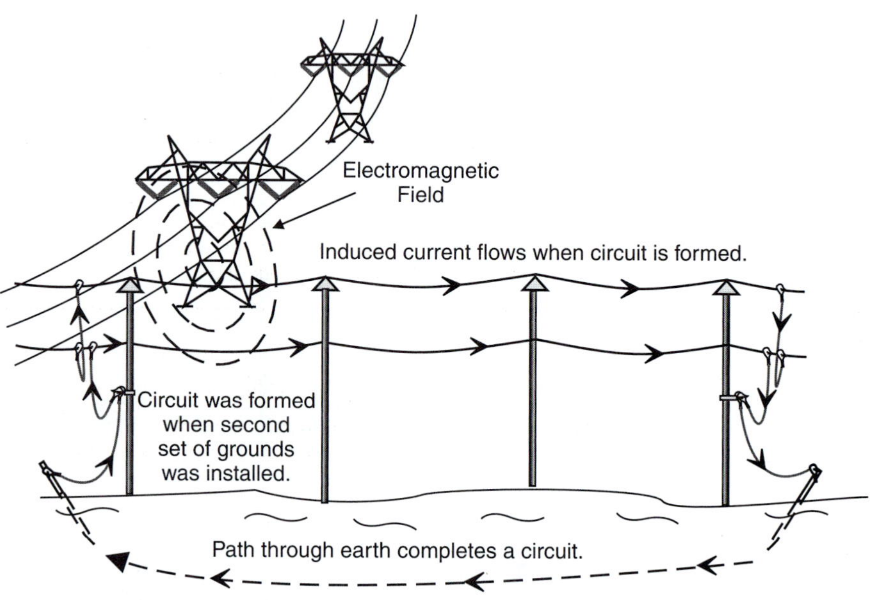

Figure 9–2   Protection from magnetic-field induction.

Sequential grounding is required to control the high-inductive circulating current in extreme cases, such as on a double-circuit 500kV. Substation grounding switches at each end of the circuit are applied to the line first. Circuiting current will flow between the two grounding switches. When grounds are installed at the point of work, the current flow in the grounds is much reduced, as seen in Figure 9–3.

(continued)

Control Current Flow	How

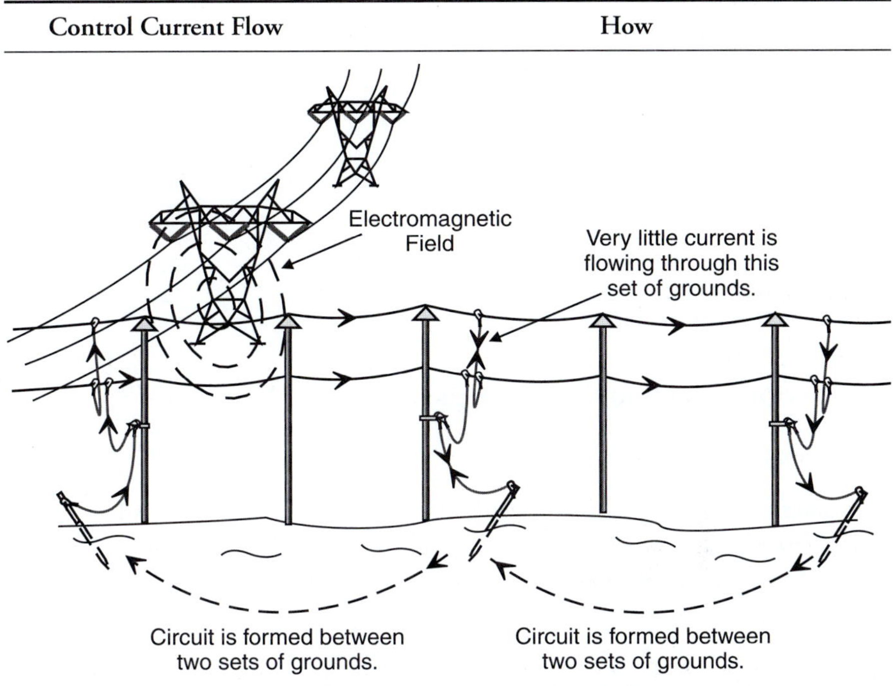

Figure 9–3 Working between bracket grounds.

Another option is to install one set of grounds only at the point of work. Any additional installation of grounds would create a circuit and cause high-circulating current flow.

A major arc can be expected when a second set of grounds is being installed or removed.

The Continuous Current Rating column in Table 9–1 shows the amount of current a ground cable can carry continuously.

The ratings in Table 9–1 are intended to give a line worker an approximation. The Fault Current column assumes that the source breaker (fuse, etc.) will trip out within 30 cycles (which is slow). Grounding cable would not be able to withstand a fault current continually, but then neither would the circuit conductors, connectors, and protective devices.

Note: When two sets of grounds are used in parallel, the fault current-carrying rating is not doubled. Parallel grounds are de-rated because any difference in resistance will cause one cable to carry more current than the other.

The Continuous Current Rating column applies to the continuous inductive current that grounding cable may have to carry.

TABLE 9–1   Grounding Cable Current-Carrying Capacity

Grounding Cable Size	Fault Current Rating	Continuous Current Rating
#2	10 kA	200 A
2/0	20 kA	300 A
2 × 2/0	36 kA	—
3/0	25 kA	350 A
4/0	30 kA	400 A
2 × 4/0	54 kA	—

9.2.2   Apply the Bonding Principle to Control Voltage

The Bonding Principle

The Bonding Principle: Bonds must be installed so that a worker is kept in an *equipotential zone.* A worker must not be able to bridge between a grounded circuit and any unbonded structure, vehicle, boom, wire, or other object not tied into the bonded network.

Control Voltage	How
1. Provide equipotential bonding to control voltage.	When protective grounds are installed, ensure that everything a person is likely to touch will be at the same potential while working on the circuit. If there is little or no potential difference across a person's body, no current can flow. When working on a pole or steel structure, a powerline worker is a parallel path to ground. Attaching the grounds to the pole, as shown in Figure 9–4, keeps the worker in an equipotential zone.

(continued)

Control Voltage	**How**

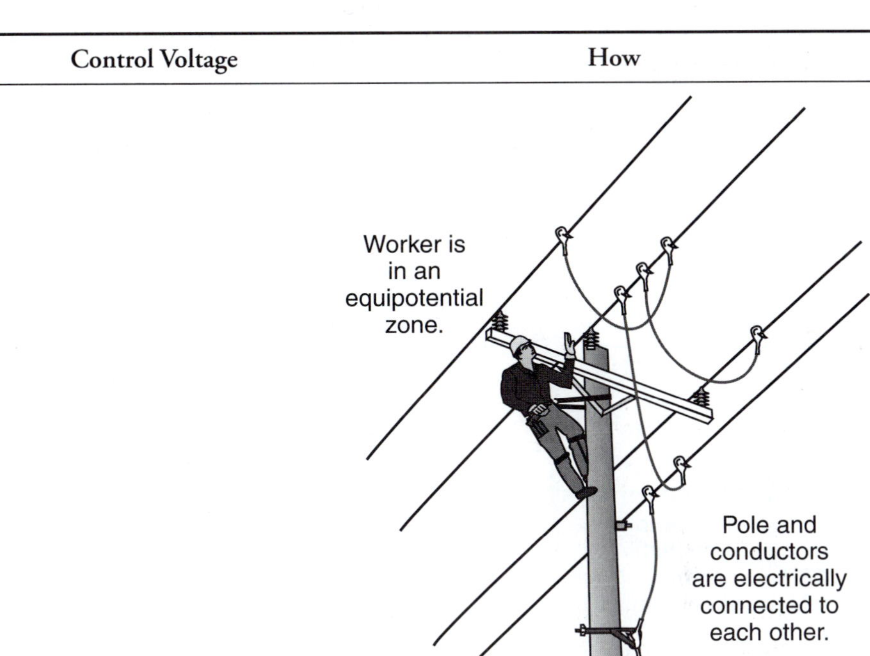

Figure 9–4 Worker in an equipotential zone.

Control Voltage	**How**
2. Bond to control voltage rise because of induction or reengergization.	With protective grounds already installed, the voltage on a circuit that is accidentally reenergized will rise in the complete circuit: the grounds, the neutral, and the earth where a ground rod is installed. If a vehicle is connected to the grounding setup, the voltage will rise on it as well.

Any component not connected electrically to the bonded network will be at a different potential, until the circuit trips out or the source of induction is removed. A person working on a pole or structure that is not bonded to the grounds will be a parallel path to ground through that pole or structure.

At the site where a set of protective grounds is installed, the phase and the neutral or the phase and steel structure are tied together and will remain at an equal equipotential when exposed to induction or accidental reenergization. When

(continued)

Control Voltage	**How**

working farther away from the installed set of grounds, the potential on the phase will remain high while the potential on a multigrounded neutral or a remote structure will drop quickly, as illustrated in Figure 9–5.

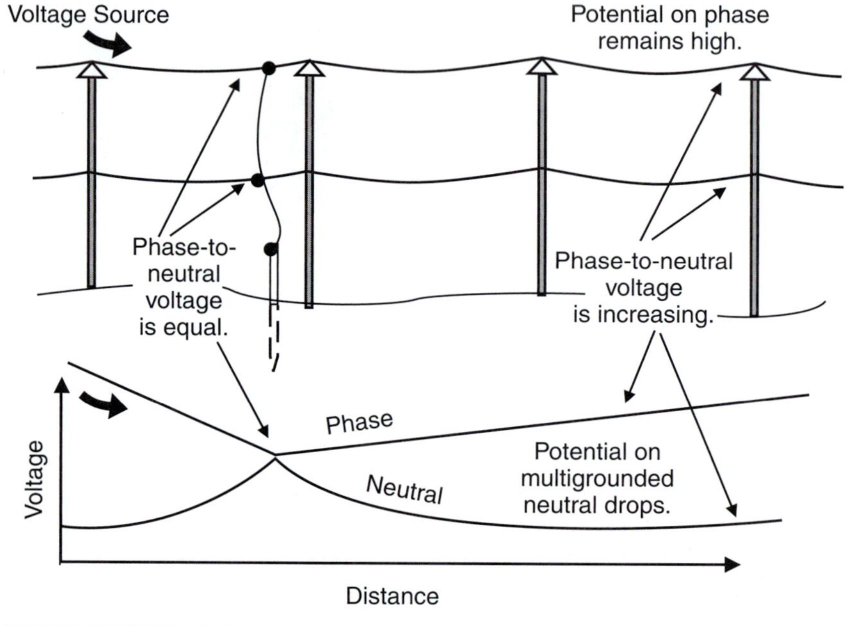

Figure 9–5 Point-of-work grounds.

The difference of potential between the phase and a neutral or structure can become dangerous when working more than 300 ft. (100 m) from the installed set of grounds. A bond can be installed even when working a span away, as illustrated in Figure 9–6.

(continued)

Control Voltage	How

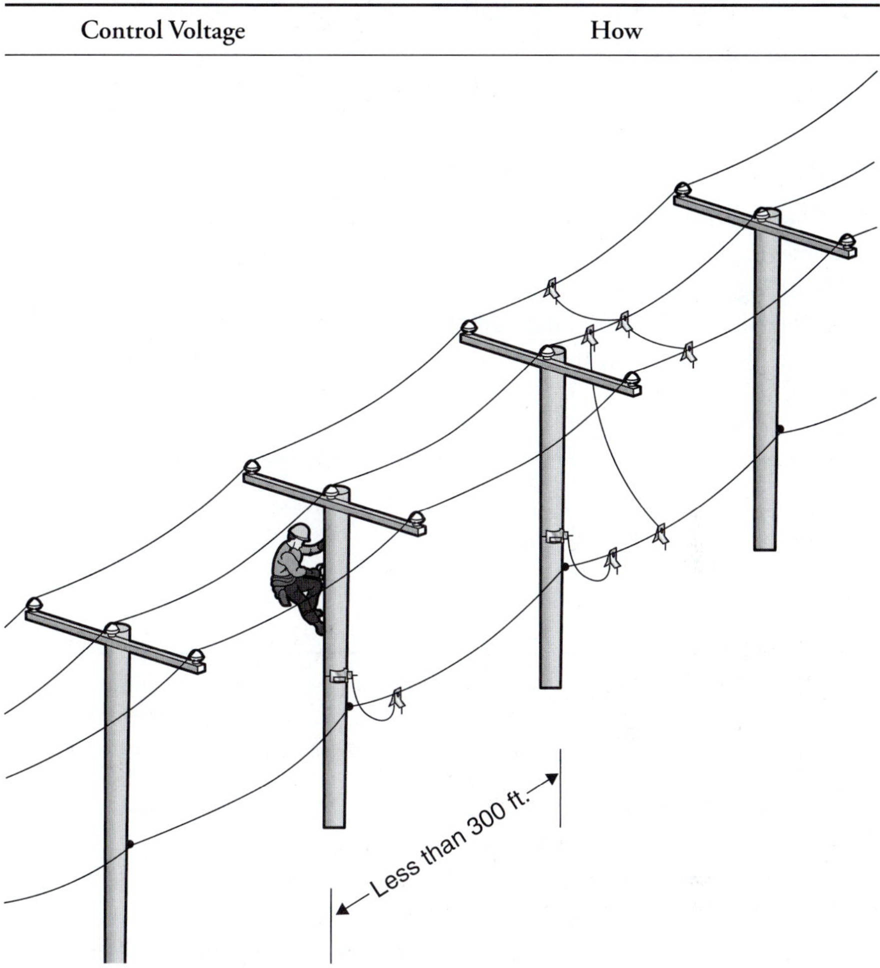

Figure 9–6   Bonding at another pole.

3. Bond the structure for equipotential grounding.	When an isolated circuit becomes accidentally reenergized or is exposed to high induction, the potential at the work site will rise and be at a different potential than any other objects not bonded to the grounded conductors. The structure will be a parallel path to ground. To eliminate the potential difference between the structure and the circuit, the structure is bonded to the circuit.
	Making a bond between the grounded conductors and a wood pole will not be a typical electrical connection. There are various methods, such as to install a pole band or connect to a bolt through the pole; however, use the method specified by your employer.

9.3 Procedures for Applying Protective Grounds on Overhead Lines

Every utility/employer has some very rigid concept about how circuits are to be grounded. The grounding procedures described in this section would be superseded by an employer's procedure.

9.3.1 Distribution Equipotential Grounding/Bonding Procedure

Equipotential grounding (also called *single-point grounding*) refers to grounding and bonding to a single point at a work site so that everything to be worked on or touched is tied together. Therefore, everything a line worker is likely to touch will have the same potential and no current flow, even during an accidental reenergization.

Step	Action	Details
1	Arrange for electrical isolation of the circuit.	Use a lockout/tagout procedure to get a guarantee that the circuit is isolated and will remain isolated. Apply tags and/or locks on isolation points.
2	Choose suitable grounding location(s).	Install a ground at the point of work. Grounding on the structure being worked on or grounding within 300 ft. (100 m) of the work location is considered grounding at the point of work. **Note:** When personal protective grounds are not installed on the structure being worked on, a bond must be installed from the structure to one of the conductors (neutral), as shown in Figure 9–6.
3	Determine the proper grounding cable size.	The grounding cable and clamps must be large enough to carry the short-circuit current available at the work location to ground and stay intact until the circuit protection opens. The level of short-circuit current available will be dependent on the ability of the system to feed the fault. If the work site is close to a substation and the circuit conductor is large, the circuit can deliver a large amount of current. A utility planning engineer would have specified the available fault current for the circuit to be worked on. Many distribution

		utilities specify a minimum cable size for the whole utility and, perhaps, designate some specific locations for larger ground cables.
4	Choose the most effective ground electrode.	On a wye system, the system neutral is the best electrode. Some utilities/employers may insist on an additional cable going to a ground rod in earth. On a delta circuit, the ground electrode will be the best electrode found at the earth level—usually an anchor or a ground probe.
5	Establish a bonded work zone at the work site.	Install bonds between the ground electrode and the pole. If the pole is wood, this means making an electrical connection between wood and the ground electrode: • Install a ground support assembly around the pole, *or* • Install a jumper from the neutral to any bolt through the pole. Tightening the bolt first will improve the electrical connection. To keep everything in the work zone bonded, any wires or winch cables brought into the work zone must be bonded to the same ground electrode. Any down guys without an insulated guy breaker or insulated rod will be a remote ground that must be bonded to the protective grounds or ground electrode. Workers on the ground who are handling conductors are outside of the bonded zone and must use rubber gloves or, in more hazardous situations, ground-gradient mats.
6	Test the circuit to verify that it is electrically isolated.	Use an approved potential tester to verify that the circuit is isolated. Buzzing or teasing is not a reliable method. The hook of a clamp stick isolation (shotgun) is the only metal in the head of the stick, and on distribution voltages it is not enough metal for a reliable indication. If more metal is used, buzzing still does not adequately distinguish between induction and an actual live line.

(continued)

9.3.1 Distribution Equipotential Grounding/Bonding Procedure
(continued)

Step	Action	Details
7	Clean the conductor.	Using a wire brush on a live-line tool, brush the conductor where the ground clamp is to be installed. Some employers will allow the use of self-cleaning clamps.
8	Install the ground clamps on the phase conductors.	Ensure that all personnel are either completely in or out of the bonded work zone before installing the protective grounds. Use a live-line tool (shotgun) that is at least 8 ft. (2.5 m) and wear eye protection. Some utilities/employers will also insist on rubber gloves. Using grounding jumpers with a support stud arrangement allows you to keep a safe distance away from the cable while installing the ground. • On a *wye system,* some employers will require a cable connection to an electrode at the earth level. On a *delta system,* one cable will be connected from the ground electrode directly on the conductor. • Using a grounding jumper cable, put one grounding jumper clamp on the neutral. • Then, in one firm movement, apply the other end to the center phase. Going to the center phase allows the grounds to be in a T formation. • Install the remaining ground sets from the center phase to each outside phase. **Note:** If a ground is mistakenly installed on live circuit, *do not remove* the ground clamp. Removing the ground clamp could draw a very long and hazardous arc. Let the circuit protection take out the circuit.
9	Remove personal protective grounds in reverse order.	To avoid accidently removing the wrong clamp do not remove any clamps from a ground rod in earth until all the ground clamps have been removed aloft. Use a clamp stick and personal protective equipment when removing the ground clamps. In high-induction areas, removing the second-to-last ground will interrupt a circulating current flow.

9.3.2 Distribution Multipoint (Bracket) Grounding Procedure

Bracket grounding refers to working between grounds on a circuit where the grounds can be 2 miles (3 km) apart, typically at each potential source.

- Bracket grounding does not adequately protect a worker from accidental reenergization; therefore, it must be seen as an exception to equipotential grounding.

- Extra steps, such as "double isolation," should be put in effect to guarantee that no accidental reenergization can occur.

- Every time a bracket grounding procedure is used, a written job plan for grounding, approved by a supervisor, should be created. The grounding plan should identify all potential sources of reenergization and the control barriers that are put in place.

- Bracket grounding *will protect* a worker from induction and backfeed from small portable generators.

- Bracket grounding should be used only where it is very impractical to ground at the work site *and* where accidental reenergization is impossible. An example of where it is impractical to apply equipotential grounding at each point of work is a stringing operation (during a large planned outage) where existing wire is being removed and new wire is being put back up in its place on many structures.

Step	Action	Details
1	Arrange for electrical isolation of the circuit.	Use a lockout/tagout procedure to get a guarantee that the circuit is isolated and will remain isolated. Apply tags and/or locks on isolation points.
2	Use double isolation to ensure that the line cannot be reenergized. Check for additional reenergization hazards.	<ul><li>Remove risers at switchgear.</li><li>Change out isolation points with suspect insulation: flashed, broken, and suspect brands or type.</li><li>Isolate suspect switchgear by removing the risers. Disable and lock out SCADA operated switchgear.</li><li>Remove any potential for the work operation to cause contact with overbuild, underbuild, or crossing circuits.</li><li>Check physical condition of pins, insulators, ties, crossarms, etc., of overbuild and crossing circuits.</li></ul>

(continued)

9.3.2 Distribution Multipoint (Bracket) Grounding Procedure
(continued)

Step	Action	Details
3	Choose the suitable grounding locations.	Grounds must be placed on all sides of the work zone not more than 2 miles (3 km) apart.
4	Determine the proper cable size.	The grounding cable and clamp must be large enough to carry to ground the short-circuit current available at the work location and to stay intact until the circuit protection opens. Technically, bracket grounding is used only where it is virtually impossible for the circuit to be reenergized; however, it is still valid to use the ground cable size meant for the system in that location.
5	Choose the most effective ground electrode.	On a *wye system,* the system neutral is the best electrode. Some utilities/employers may insist on an additional cable going to a ground rod in earth. On a *delta circuit,* the ground electrode will be the best electrode found at the earth level, usually an anchor or a ground probe.
6	Test the circuit to be grounded.	Use an approved potential tester to verify that the circuit is isolated. Neither buzzing nor teasing is a reliable method.
7	Clean the conductor.	Using a wire brush on a live-line tool, brush the conductor where the ground clamp is to be installed. Some utilities/employers will allow the use of self-cleaning clamps.
8	Install the ground clamps on the phase conductors.	Ensure that all personnel are either completely in or out of the bonded work. Use a live-line tool (shotgun) that is at least 8 ft. (2.5 m) and wear eye protection. Some utilities/employers will also insist on rubber gloves. Put one grounding jumper clamp on the neutral and then, in one firm movement, apply the other end to the closest phase conductor. Install ground sets from the grounded conductor (or neutral) to the other conductors.

(continued)

Step	Action	Details
9	For additional protection, establish a bonded zone at the work site.	When working between bracket grounds, the grounds will reduce potential on the conductors and trip out a circuit if it should become reenergized. However, in high-induction areas the grounds are not on the structure being worked on, so a bond must be installed from the structure to one conductor (neutral). In a high-induction area, the induction on the bracket grounded circuit will be low; the induction on work equipment, and on any wires or winch cables brought into the work zone, could be high and at a different potential than the grounded circuit.
10	Remove grounds in reverse order.	To avoid accidentally removing a wrong clamp from the ground probe, do not remove any clamps until all the ground clamps have been removed aloft. Use a clamp stick and personal protective equipment when removing grounds.

9.3.3 Transmission Equipotential Grounding/Bonding Procedure

Step	Action	Details
1	Arrange for electrical isolation of the circuit.	Use a lockout/tagout procedure to get a guarantee that the line will be isolated and remain isolated. Verify that the control center has isolated the proper circuit and opened and locked/tagged the correct switchgear by making reference to an operating drawing. If grounding switches are on the circuit, ensure that they are closed.
2	Choose suitable grounding location(s).	Grounds must be installed in a set sequence to control current where induction from neighboring circuits is high. Substation grounding switches at each end of a circuit are applied to the line first. Circuiting current will flow between the two grounding switches when grounds are installed at the point of work.

(continued)

9.3.3 Transmission Equipotential Grounding/Bonding Procedure
(continued)

Step	Action	Details
		The point-of-work grounds are installed last. A much reduced current flow is in the grounds, as seen in Figure 9–3. Another option in high-induction areas is to install one set of grounds at the point of work. Any additional installation of grounds would create a circuit and cause high circulating current flow.
		Where induction is not a major concern, each crew can install grounds at the point of work. The second set of grounds will create a circuit, and some current flow may still be set up.
3	Determine the proper grounding cable size.	The grounding cable and clamp must be large enough to carry the fault current available at the work location and stay intact until the circuit protection opens. The utility is responsible for specifying the available fault current on the circuit and the size of the appropriate ground cable and clamps.
		Note: If parallel grounds are needed, the first set of grounds installed on a circuit is, supposedly, not able to carry the fault current and will fail if the circuit being grounded is still hot.
4	Choose the most effective ground electrode.	On *steel poles and steel-lattice structures,* install a ground stud on the structure. With the shield wire properly bonded to the steel structure, the structure becomes a good ground electrode. The structure is also well bonded to the conductor after a ground is installed.
		On *wood pole lines,* use an anchor or a ground probe and install a ground clamp on the shield wire.
5	Establish a bonded work zone at the work site.	On *wood poles,* install a ground support assembly on the pole and bond it to the down leads, guys, and ground probes.
		Steel structures provide a bonded area to workers aloft.
		For *both wood pole and steel work,* install bonds on any wires, booms, or winch cables brought into the work zone.

(continued)

9.3.3 Transmission Equipotential Grounding/Bonding Procedure
(continued)

Step	Action	Details
		Ensure that workers *on the ground* are protected by using ground-gradient mats or rubber gloves, as applicable.
6	Test the circuit to verify that it is electrically isolated.	Use an approved potential tester. Teasing/buzzing is not a reliable method for testing isolation because it does not adequately distinguish between induction and an actual live line.
7	In addition to installing single-point grounds at the work site, install bracket grounds where necessary.	Install bracket grounds on each side of the work site under the following conditions: • The work site is in a high-induction area. • Poor soil conditions at the work site reduce the effectiveness of the installed ground probe. Bracket grounds will reduce the voltage of the bonded work zone in relation to remote ground and will reduce the current in the on-site protective grounds.
8	Clean the conductor.	Using a wire brush on a live-line tool, brush the conductor where the ground clamp is to be installed. Some employers will allow the use of self-cleaning clamps.
9	Install the ground clamps on the phase conductors.	• Ensure that all personnel are either completely in or out of the bonded work zone before installing the protective grounds. • Use live-line tools and eye protection. • Ground the closest phase conductor first. • On *wood poles,* install grounds on all three phases. On *steel structures,* some utilities allow only the phase being worked on to be grounded. • If conductors must be dropped to the ground, three separate grounds from a common electrode will need to be installed. See Figure 9–7 for an illustration of equipotential grounding on an H-frame transmission line. Note the **T** formation of the grounds.

(continued)

9.3.3 Transmission Equipotential Grounding/Bonding Procedure
(continued)

Step	Action	Details 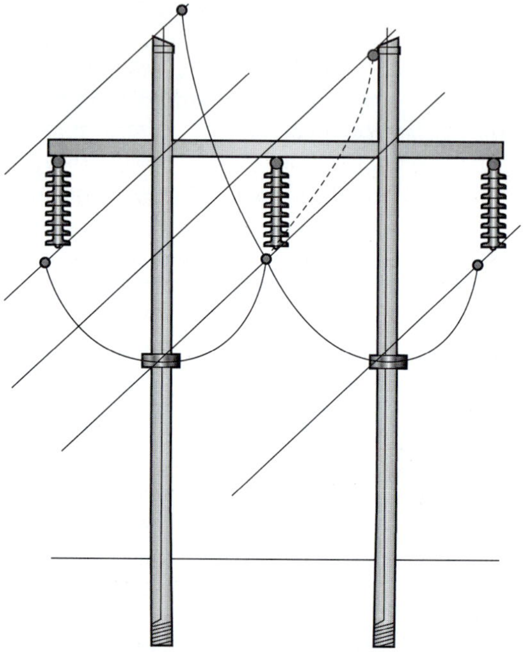

Figure 9–7   Typical transmission line grounding.

Note: If a ground is mistakenly installed on live circuit, *do not remove* the ground clamp. Removing the ground clamp could draw a very long and hazardous arc. Let the circuit protection take out the circuit.

Step	Action	Details
10	Remove personal protective grounds in reverse order.	• Remove the clamp from the shield wire after the phase clamps are removed. There could be a high-potential difference between the grounded phases and the shield wire. • To avoid accidently removing a wrong clamp from the ground probe, do not remove any clamps until all the ground clamps have been removed aloft. • Use a clamp stick and personal protective equipment when removing grounds. • When the second-to-last ground is removed, an electrical circuit is being opened and an arc could occur.

(continued)

9.3.4 Transmission Bracket Grounding Procedure

At times, a transmission-line bracket grounding procedure can be very practical, such as a project involving the setting of poles in deenergized circuit. However, bracket grounding *does not adequately protect* a worker from accidental reenergization but *will protect* a worker from induction. If bracket grounding is used, it must be considered as an exception to the single-point grounding/bonding that is planned, documented, and approved by a utility/employer.

Bracket grounding should be used only where it is impractical to ground at the work site and where accidental reenergization is not possible. A written job plan for grounding should include any extra steps taken to disable any sources of renergization, including removing a span, removing loops at a switch, and so on.

9.4 Personal Protective Grounding on Underground Cable

9.4.1 Reasons to Ground Underground Cable

Purpose	Details
1. Install personal protective grounds to prove isolation.	Underground cable is difficult to trace physically and cutting into live cable has not been uncommon. It is critical to test and ground at the point of work. Test and ground where there is access to the cable, and spike the cable where there is no access.
2. Install personal protective grounds for protection from accidental reenergization.	When working on a grounded cable, a set of protective grounds must always be in place for protection from accidental reenergization.
3. Install personal protective grounds to provide protection from induction.	Underground cable is a capacitor that can maintain a charge for a long time. A protective ground must be installed and maintained to drain such charges.

9.4.2 Apply the Grounding Principle

A protective ground, sized for the available fault current, is needed to trip out the protective switchgear in case of accidently grounding a live circuit or in case of reenergization. Apply grounds to the cable wherever it is possible to get access—for example, at risers, at switching cabinets, at transformers, and at live-front switchgear cabinets. Spike the cable when needed to prove isolation.

9.4.3 Apply the Bonding Principle

If a cable has personal protective grounds installed and is accidently reenergized, the protective grounds will trip out the circuit. However, the grounds are not like a gate where all the current is shunted to ground. There will be a momentary rise in potential on the core conductor, the sheath, the grounding network in a cabinet, and everything bonded to the protective grounds. Anyone touching or near the cable, transformer, or switching cabinet (kiosk) will be a parallel path to a remote ground and subject to a lethal electrical burn.

An equipotential zone must be created so that a worker is totally bonded to the protective grounds. A ground-gradient mat bonded to the protective grounds will keep a worker within an equipotential zone. The mat serves the same function as bonding a structure to the protective grounds in an overhead lines situation. When working on a cable in a trench, work from a ground-gradient mat bonded to the sheath/concentric neutral.

If a cable is being cut in two, a high potential is possible between the two ends of the cable, even with protective grounds installed at each end. A jumper should be installed on the sheath to bond the two ends before cutting to keep continuity between the grounds.

During a cutting or splicing operation, when no exposed cable or sheath is available to which to bond, rubber gloves and grounds at the terminals will be the only protection available. Exposure time to these conditions should be planned and kept as short as possible.

9.4.4 Cable Grounding Locations

Apply grounds to the cable wherever it is possible to get access—for example, at risers, at switching cabinets, at transformers, and at live-front switchgear cabinets.

Grounding at Risers

Conventional protective ground sets can be applied at a riser pole (also called a *dip pole* or *transition pole*), as long as enough conductor is exposed somewhere on the structure. One method of allowing a protective ground to be connected to the bottom of a cutout is to use a clamp similar to the device shown in Figure 9–8.

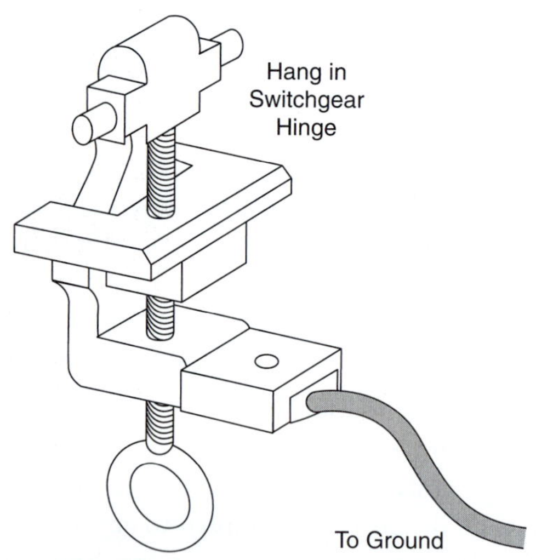

Figure 9–8   Ground clamp attachment.

Grounding with a Load-Break Elbow

A load-break elbow plugged into a transformer or other dead-front equipment can be grounded by removing the elbow from the transformer bushing and inserting it into a grounded bushing. A portable feed-through (as seen in Figure 9–9) can serve as the grounded bushing.

Grounding at a Live-Front Transformer or Switchgear

On live-front equipment, the load side of switchgear is exposed and somewhat accessible to protective grounds that have clamps suited to the equipment to be grounded. Isolating the cable to be grounded requires opening the appropriate switches at both cable terminals. One way to ground is to remove the fuse barrel (door) and place a ground at the load side of the switch using the proper clamp, as shown in Figure 9–8. Ball-and-socket connections also are available for grounding at a switch.

Grounding Between Terminals

To work on a cable between terminals, first each end of the cable must have grounds installed. At the work location, the cable is then spiked to positively identify the cable as isolated and jumper the sheath before cutting through the cable. A spiking tool can be driven into the cable, which will short it out. If the cable does not have a grounded concentric neutral, the "ground lug" on the tool has to be connected to a driven ground rod or any other good ground electrode.

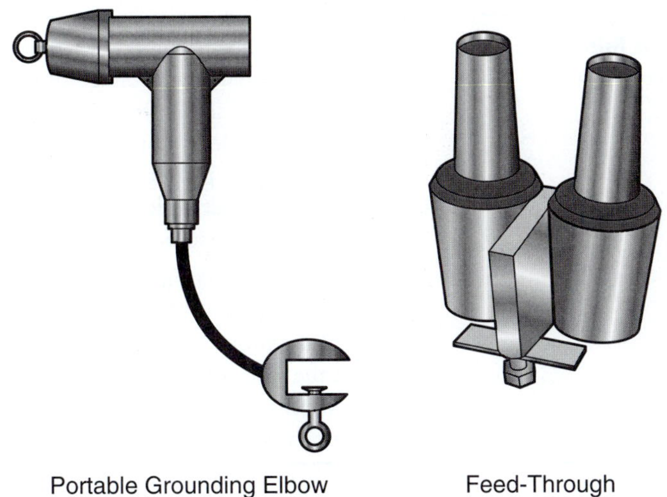

Figure 9–9 A grounding elbow and a feed-through.

9.5 Specific Grounding Hazards

Hazard	Procedural Control
1. Installing grounds on a live circuit	In addition to getting an *isolation guarantee* on a circuit, a line worker must identify the correct circuit, conductor, or cable in the field with a potential tester suitable for the voltage rating of the circuit. Teasing (buzzing) a conductor with the metal head of a live-line tool is not an effective method of checking for isolation. *Applying protective grounds is the ultimate test for isolation.* Figure 9–10 shows one of the three or four potential testers available to test for potential on a high-voltage circuit.

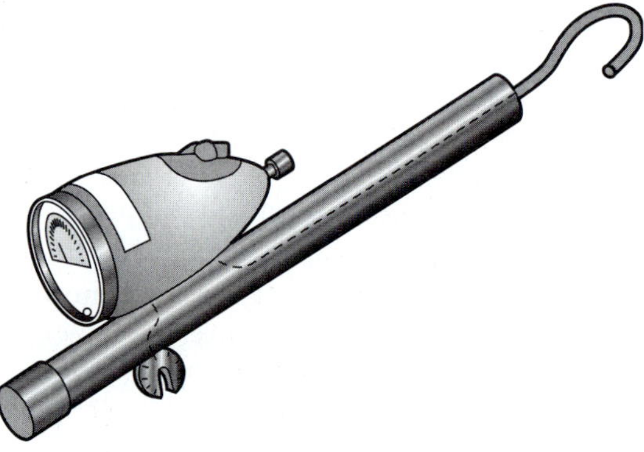

Figure 9–10   A voltage detector.

Hazard	Procedural Control
2. Cutting or joining a grounded conductor	If a current flow is interrupted by a break in a circuit, an immediate high voltage appears across the break. Even if a set of protective grounds is installed on each side of a break in the conductor, it is still likely that a potential difference might occur across the break, unless the two grounds are connected together electrically. Use a live-line tool to install a jumper across an open point in a circuit. When planning the location of the protective grounds, any switchgear or fuse in the work zone that could operate during reenergization should be treated as a potential break or open point.

(continued)

9.5 Specific Grounding Hazards *(continued)*

Hazard	Procedural Control
3. Accidental reenergization or induction will create touch and step potentials around any ground probes, vehicles, wire, etc., bonded to the protective grounds.	Workers on the ground are very vulnerable unless they are in the bonded zone and must continuously apply the bonding principle for protection from electrical shock. Applications of the bonding principle while working on the ground include, but are not limited to, the following: • Stay clear of everything attached to the bonded network. • When operating a boom, stay on the truck platform. • If tools are required from a bonded vehicle, use rubber gloves. • If a ground wire must be sent aloft, use a ground-gradient mat. In a station yard, the station ground network and the graveled yard provide some protection from ground gradients.
4. Wood pole work in poor soil or in high induction areas	When working on a three-wire system, especially on transmission lines, protective grounds will not necessarily reduce the voltage of the grounded conductors to acceptable levels in relation to a remote ground. A truck boom or unbonded down guy being sent aloft could become a source of a remote ground if not bonded to the protective grounding setup. When available, grounding to the shield wire (static wire, ground wire, or counterpoise) will lower the voltage on the bonded system. The shield wire can be the best ground electrode available at the work site because it is multigrounded and a direct path to the station ground. In addition to installing grounds at the work site, installing bracket grounds (grounds somewhere on each side of the work site) will lower the voltage at the work site. The bracket grounds can be installed in any convenient location that has a good ground electrode.

(continued)

9.5 Specific Grounding Hazards *(continued)*

Hazard	Procedural Control
5. Grounding gradients when only one phase has a protective ground installed	Grounding all three phases results in fast clearing in the case of reenergization. When the three phases are tied together, less current flows through the ground cable and less ground current flows to earth. An additional benefit is a reduced level of touch and step potentials for workers on the ground. On steel transmission lines, some utilities will allow grounding only of the phase being worked on. The steel provides a good ground electrode, and grounding only one phase will trip the three-phase circuit breaker during an accidental reenergization. The workers on the steel structure will be working in a bonded area, but workers on the ground should stay clear of the tower.
6. Protective grounds whipping during a fault	During a high-fault current condition, the portable ground cable can whip violently. The ground cable can also burn off if it is wrapped around a steel structure or is left coiled. The huge magnetic field around the ground cable when it is carrying a high-fault current results in major mechanical forces exerted on the cable. Use a rope to tie off grounds where workers could be exposed to the whipping action. Keep grounds as short as practical, and lay out the ground so it is not coiled or wrapped around steel.
7. Protective grounding for backfeed from a portable generator	Backfeed from a generator that was improperly connected at the customer can feed back through the secondary of the transformer and create a high potential on the primary side. When working on an isolated circuit, protective grounds on the circuit will reduce the voltage due to backfeed from a portable generator to an acceptable level. The transformer impedance and the resistance of the conductor between the grounds and the generator will probably not create a fault current large enough to cause the generator to overload.

(continued)

9.5 Specific Grounding Hazards *(continued)*

Hazard	Procedural Control
	As long as the generator keeps running, the voltage on a conductor with protective grounds installed will be a current flow in the conductor. Because current flow in a conductor is always a possibility, the conductor should not be cut or joined without first installing a jumper. The magnitude of the voltage and current from a large industrial or commercial generator would cause a higher voltage and current to appear on grounded conductors at the work site, but because these generators are more likely to be installed and inspected properly, backfeed is very unlikely. The hazard from backfeed is from the unknown portable generator.
8. Working from a transformer pole or lateral tap pole	If work is to be done on a transformer pole, and a source of unknown secondary backfeed is present, the protective grounds on the primary will protect a powerline worker as long as the transformer fuse is intact. However, it is against protective grounding principles to rely on a fuse for continuity of grounds. The fuse could blow or the cutout could fall open, leaving the high-voltage transformer bushing at full-line voltage. The secondary transformer leads should be removed or should be shorted out with some heavy battery-boosting cables before working on the power conductors.
9. "It is not dead if it isn't grounded" is not entirely true.	After protective grounds are installed, voltage relative to a remote ground can still be present, as can current flow in the conductors and ground cable. • Induction can cause the potential on the conductors to be different from a remote ground introduced into the work site, such as a truck winch or a down guy without an insulated breaker.

(continued)

9.5 Specific Grounding Hazards *(continued)*

Hazard	Procedural Control
	• Induction can cause current flow when a second set of grounds installed on a line creates a circuit.
	Personal protective grounds do lower the voltage to a safe level and control any current flow. The saying "It is not dead if it isn't grounded" is not entirely true. A more correct saying is "It is only safe to work on a circuit after personal protective grounds are installed properly."

Stringing Overhead and Pulling Underground

10.1 Stringing Overhead Conductors

10.1.1 Stringing or Removing Distribution Conductors Near a Live Circuit

Hazards	Controls
While stringing a conductor in the vicinity of a live circuit, the conductor could make an inadvertent contact.	Tension stringing techniques should be used when stringing over or at the same level and on the same side of the structure as a live circuit. If slack stringing underneath a live circuit, use tie-down ropes midspan to prevent a snagged conductor from whipping up and making contact with the live circuit.
Workers on the ground are exposed to possible electric shock while stringing in the vicinity of a live circuit.	If slack stringing, the reel stands should be grounded and put on a ground-gradient matting or on a grounded trailer. Anyone tending the reels must remain on the ground-gradient matting or trailer. An operator of a reel trailer, tensioner, or puller must stay on the operating platform or on a ground-gradient mat that is bonded to the machine. The conductor must be held with grips and protective grounds installed before changing reels or doing other work at the tensioner or puller end.
A conductor makes contact with a live circuit while a worker is tying or clamping it to an insulator. The live circuit may not trip out when a conductor contacts the live circuit.	Rubber gloves must be worn by anyone handling the conductor unless they are in a bonded zone. A stringing procedure and precautions taken should recognize that a conductor can become alive and that it may not trip out the circuit. Ground the stringing blocks to the system neutral at regular intervals. Use traveling grounds at the reel stand or tensioner end.

10.1.2 Stringing Across Roadways

Reduce the risk of a vehicle making contact with conductors being strung across a roadway by considering the following:

- Road-crossing spans are often longer than other spans in the line being strung. A lot of conductor will gather in an extra-long span and cause excessive sag. Consider stringing the road-crossing span by itself instead of making it just another span in a long pull.

- On high-volume roads, consider using road authorities and their equipment to channel and control traffic. The presence of a highly visible police vehicle may add to the effectiveness of traffic control.

- If tension stringing, use backup barriers such as rider poles or a rope basket to catch conductors before they drop onto the road. Slow traffic by channeling vehicles into a single lane so that they can be stopped quickly if a conductor drops too low.

- If slack stringing, stop traffic before running a rope or wire across the road. Stopped traffic in each direction can serve as a barrier for traffic entering the work zone.

- A pole can break when a passing vehicle catches onto a conductor or when a running board catches on a stringing block. No one should be on a pole when the conductor is in motion. To free a running board from a stringing block, move the conductor only far enough for a person to go aloft to fix it.

10.1.3 Typical Distribution Tension Stringing Procedure

Step	Action	Details
1	Ensure adequate spacing to allow stringing a new conductor.	1. To install a new circuit above an existing circuit, higher poles must be installed. The space above the live underbuild should be at least 5 ft. (1.5 m) and ideally 10 ft. (3 m). 2. To replace existing conductors, install auxiliary arms and set out the existing live conductors. The closest conductors should be at a minimum of 3 ft. (1 m) to the side. 3. To cross over live circuits mounted on the same pole, the necessary clearance can be smaller because the live circuit can be covered up. Crossing live circuits in-span requires more clearance, but additional protection can be obtained by installing cover-up.
2	Install stringing blocks.	The *finger line* (throw line) is a small rope put through the stringing blocks and tied off down the pole. The finger line is used later to pull the pilot line or pulling rope through the stringing block. 1. Stringing blocks are normally installed when any necessary framing is done. Finger lines installed

(continued)

10.1.3 Typical Distribution Tension Stringing Procedure *(continued)*

Step	Action	Details
		through the stringing blocks during installation will save additional setup. The finger lines must be tied off above public reach and clear of live underbuild. A clean finger line, in combination with rubber gloves, will tolerate momentary brush contact with live conductors. 2. On multisheave stringing blocks, keep the finger lines in the center sheave. 3. Corner stringing blocks should, if possible, be set out so that the conductors do not have to be moved into place before sagging. 4. Install grounds, as required by your utility, on the first and last stringing block and at regular intervals. 5. **NOTE:** Friction from stringing blocks is a large factor in determining the pulling tension. *Pulling Tension = Line Tension + (% Friction of Block × Line Tension × Number of Structures)* For example, pulling tension for a 20-span pull, with a 3% friction of blocks and line tension of 500 pounds = 500 + (.03 × 500 × 20) = 800 pounds.
3	Set up the puller.	1. To reduce the down weight on a structure, keep the puller back at a distance of at least one-third of the span length. 2. Set up the puller so that it is in line with the pull or so that the installation of snatch blocks will allow the pull to be in line. 3. Ensure that the operator and anyone else near the machine will be on an operating platform or on a ground-gradient mat bonded to the puller. 4. Install a barrier around the puller to keep away people during the stringing operation. The electrical hazard at the puller increases when the conductor comes into the puller. 5. Use a reinforced steel reel made specifically for winding up the pulling rope under tension. Other reels can be crushed by the elastic properties of the rope.

(continued)

10.1.3 Typical Distribution Tension Stringing Procedure *(continued)*

Step	Action	Details
		6. Have protective grounds ready to install on the conductor when it comes to the puller.
4	Set up the tensioner.	The *tensioner* is the reel carrier for the conductor and holds tension on the conductors by pulling back the conductors hydraulically or by applying brakes. 1. Set up the tensioners so that the conductors will be in line with the first pole or structure. 2. Place ground-gradient mats so that the machine operator and people involved in changing reels are working from the mat bonded to the tensioner. 3. Install a barrier around the tension machine(s) to keep away people during the stringing operations. 4. Have protective grounds ready to install on the conductors when the pull is stopped. For additional grounding during stringing, install traveling grounds. 5. Wooden conductor reels can collapse and cause a sudden loss of tension. It is generally acceptable to tension directly from a wooden or metal reel at tensions less than 1,000 lbs. (450 kg). Use a metal axle all the way through a wooden reel to prevent a total collapse. Use a bull-wheel tensioner for higher tensions.
5	String in the pilot line.	A *pilot line* is a small rope strung from the tensioner, through the stringing blocks to the puller, and used to pull back the heavier pulling rope. (**NOTE:** The pulling rope may be pulled out directly without the use of a pilot line.) If stringing wire through multisheave stringing block (for example, three sheaves), only one pilot line is needed to pull in one pull rope. 1. As the pilot line is strung and fed through the stringing blocks, it will likely make occasional contact with any live underbuild. The first 150 ft. (50 m) of the pilot line should consist of live-line rope or clean, dry, synthetic rope. Rubber gloves must be worn by the handlers of the rope. 2. Guard all road crossings during the stringing of the pilot line.

(continued)

10.1.3 Typical Distribution Tension Stringing Procedure *(continued)*

Step	Action	Details
6	Pull the pulling rope back to the tensioner with the pilot line.	The *pulling rope* is the rope used to pull in the conductor. It must be strong enough to pull in the conductor, have very little stretch, and be clean enough to withstand brush contact with live conductors. 1. Connect the pilot line to the pulling rope. 2. Using a tensioner, pull the pilot line from the puller back to the tensioner.
7	Pull in the conductor with the pulling rope.	1. Use a minimum length of pulling rope to make the pull. Work as close to the drum core as possible because, unless it is a bull-wheel puller, a large-diameter reel of pulling rope will require more pulling tension. 2. The pull rope will be connected to one conductor or to a running board that may have more than one conductor attached to it when used on a transmission line. The connection to the pulling rope is made with a swivel joint. If the swivel is not performing well, a running board will start flipping over in a span, often more than once, which creates a lot of work to unwind. 3. Establish radio communications between the puller, the tensioner, observers at road crossings, and workers following the running board or rope-to-conductor joint. If possible, use an exclusive radio channel for stringing. Only the puller can stop the pull, and only the tensioner can adjust the stringing tension or sag. 4. Apply tension at the tensioners. Hydraulic tensioners are put in the take-up mode to tension the conductors. Tensioners that use braking to hold the conductors must have brakes that are designed for continuous braking and that will not overheat and fade. 5. Pull in the pulling rope on the puller. 6. When the conductor arrives at the puller, install grounds at both ends.

(continued)

10.1.3 Typical Distribution Tension Stringing Procedure *(continued)*

Step	Action	Details
8	Problems that may be encountered while pulling in the conductor(s).	1. If the conductor sag is fluctuating and hard to control, the design of the pulling rope has too much stretch. The elasticlike action is causing the tension and sag to change constantly. Use a rope specifically designed for tension stringing. 2. On a multiconductor pull, the running board can flip over between spans even when a swivel is used. The pulling rope should be designed not to rotate. 3. If there is a long span within the pull, such as a wide road crossing or river crossing, and it is difficult to maintain the high tensions needed to maintain clearance from underbuild, that span should be strung separately. The weight of the long span will keep the sag in shorter spans excessively tight. 4. Conductor smaller than 3/0 will cut into the lower layers of a conductor reel. Unless a tensioner is equipped with a bull wheel to remove the tension from the reel, small conductor should not be strung. 5. If a tensioner is tensioning the conductor directly from a reel, the braking action must be reduced as the reel empties and the circumference gets smaller. 6. For heavy conductor on a long pull, tension can be lost when brakes heat and fade on the tensioner, so a tensioner with a hydraulic retarding system should be used. 7. If the puller seems to be pulling at a very high tension and it is not tight at the tensioner: • The stringing blocks may be too small in diameter and cause extra loss due to the bending and straightening of the conductor as it passes over each stringing block. • There may be too much friction on the stringing blocks, too many corner structures or changes in elevation for the length of the pull. Losses in efficiency are additive. Even a good stringing block with a 2% loss at each

(continued)

10.1.3 Typical Distribution Tension Stringing Procedure *(continued)*

Step	Action	Details
		suspension point can, after 15 structures, increase the total tension of the pull approximately 30%. • A flexible mesh pulling grip on the pulling rope can let go when it is not properly installed. The diameter of the rope will decrease considerably under high tension. Use a fiber or wood plug in the core of the rope to maintain the rope diameter.
9	Change conductor reels at the tensioner as required.	All workers must be completely in the bonded work zone or completely out of the bonded work zone. 1. To allow reels to be changed at the tensioner, the ground-gradient mat setup should be big enough to accommodate the work. Protective grounds are installed on the conductor and bonded to the ground-gradient mat. If a vehicle boom is used from outside the bonded zone, the vehicle must be bonded to the work zone. The boom operator must stay on the platform, and all others must stay away from the vehicle. 2. However, *if* the pull is stopped, protective grounds are installed at the tensioner end, the conductors are temporarily dead-ended in grips, and the length of the pull is patrolled for any potential electric contact hazards, *then* the reels at the tensioner could be changed with a reasonable assurance that the conductors will not become energized.
10	Free the running board or conductor when itsnags onto a stringing block.	1. While the conductor or pulling line is being pulled (in motion), no one should be aloft on a pole or structure or aloft in the vicinity (in a bucket), except during a small amount of adjustment needed to guide the stringing sock or board through a stringing sheave.

(continued)

10.1.3 Typical Distribution Tension Stringing Procedure *(continued)*

Step	Action	Details
		2. When handling conductors during a tension stringing operation, use rubber gloves or apply protective grounds and/or bonds to the conductors and structure at the point of work.
11	Sag the conductor.	1. Set the corner conductors in stringing blocks as close as possible to their final positions. 2. Sag the conductors using the tensioners, or bring the conductors in close to sag using the tensioners and finish sagging with chain hoists or a derrick winch.
12	For long pulls that require mroe than onesetup, dead-end the conductors temporarily.	1. Temporary dead ends require standard anchoring and guy steel to hold the conductor tension. The use of rope guys or unsupported pole butts can lead to sagging into live underbuild. 2. Live underbuild conductors will need cover-up and/or setting out to provide clearance for the guying. 3. The slope of the guys must be adequate to prevent unacceptable downward stresses on crossarms, insulators, and so on.
13	Tie in or clamp in conductor.	1. Use rubber gloves or apply protective grounds and/or bonds to the conductors and structure at the point of work.

10.1.4 Stringing Aerial Cable

If a messenger cable (strand) loses tension after the secondary cables are lashed to it, excessive tension will be placed on the secondary cables. It will be very difficult to get enough slack out of the secondary cables to make service connections.

String the messenger cable to the specified tensions, and pre-stress the line as required. Test the anchors to ensure they will not let go while someone is aloft.

10.1.5 Measuring Sag Using the Return-Wave Method

The *return-wave method* can also be used to check the sag on an existing line. This method is applicable regardless of the span length, tension, size, or temperature or type of conductor. Start a wave in a conductor by pulling on it with a telescopic stick or jerking down on a rope over the conductor. The wave travels to the next structure and is reflected back and forth until the wave is eventually damped out. Record the time it takes (a special stopwatch is available for this purpose) for three, five, or ten return waves and Table 10-1 can be used to find the sag.

10.1.6 Tying or Clipping in Conductor

Hand tied ties or preformed ties are used on most top-tie insulators. Utility specifications and set patterns are available regarding how to tie in a conductor. Formed wire ties are very secure and protect the conductor from broken strands due to vibration.

10.1.6.1 Formed Wire Distribution Ties

The most critical part of using formed wire ties (Hubbell's Super Top Tie, PLM's wrap lock, PLM's EZ Wrap Twin, Tycoe) is ensuring the correct tie is being used. Read the identification tag to confirm the correct size and type of conductor, type of insulator, neck size of insulator, top tie, side tie, and double tie. Ties are available for different types of insulators. Table A10–1 is an aid to choosing the correct Hubbell Super Top Tie.

Two color codes are used on distribution ties. The inside color code identifies the conductor size, and the outside color code identifies the insulator neck size.

Insulator color codes

Black   C Neck

Yellow   F Neck

Green   J Neck

TABLE **10–1**   **Return-Wave Method Chart**

Sag in Inches (cm)	Return of Wave in Seconds		
	3rd Time	*5th Time*	*10th Time*
24 (0.61)	4.2	7.0	14.1
30 (0.76)	4.7	7.9	15.8
36 (0.91)	5.2	8.6	17.3
42 (1.07)	5.6	9.3	18.7
48 (1.22)	6.0	10.0	19.9
54 (1.37)	6.3	10.6	21.1
60 (1.52)	6.7	11.1	22.3
66 (1.68)	7.0	11.7	23.4
72 (1.83)	7.3	12.2	24.4
78 (1.98)	7.6	12.7	25.4
84 (2.13)	7.9	13.2	26.2
90 (2.29)	8.2	13.7	27.3
96 (2.44)	8.5	14.1	28.2
102 (2.59)	8.7	14.5	29.1
108 (2.74)	9.0	15.0	29.9
114 (2.90)	9.2	15.4	30.7
120 (3.05)	9.5	15.8	31.5
126 (3.20)	9.7	16.2	32.3
132 (3.35)	9.9	16.5	33.1
138 (3.51)	10.1	16.9	33.8
144 (3.66)	10.4	17.3	34.5
150 (3.81)	10.6	17.6	35.2
156 (3.96)	10.8	18.0	35.9
162 (4.11)	11.0	18.3	36.6
168 (4.27)	11.2	18.7	37.3
174 (4.42)	11.4	19.0	38.0
180 (4.57)	11.6	19.3	38.6
186 (4.72)	11.8	19.6	39.2
192 (4.88)	12.0	19.9	39.9

10.1.6.2 Conductor Clamps

Ideally, you will use the suspension and trunnion clamps designed for easy
and safer access with a live-line tool. Three types of clamps have one or two
easy-to-access bolts with captured nuts that allow easy installation with a flexible wrench head (see Figure 10–1).

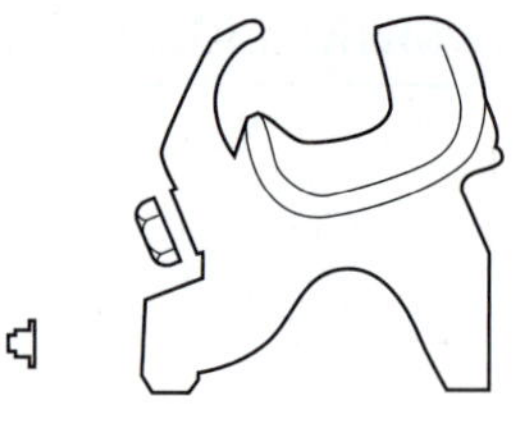

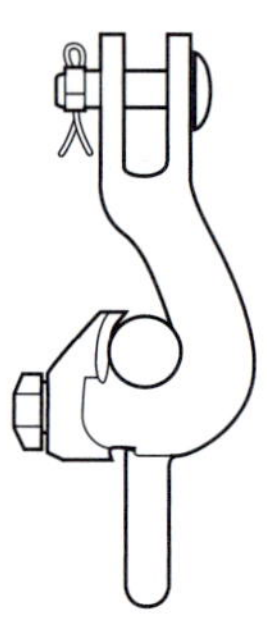

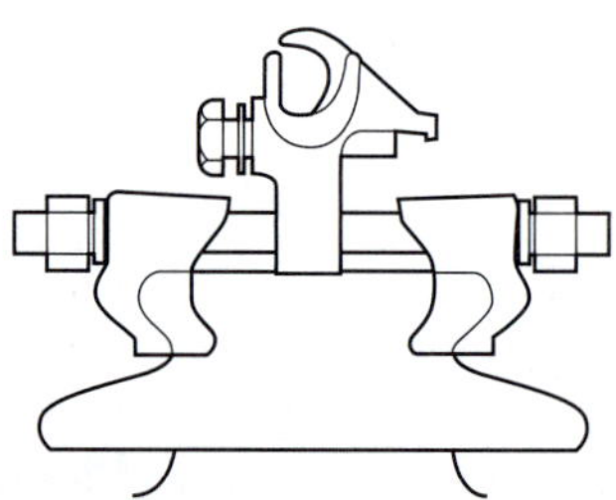

Figure 10–1 Conductor clamps.

10.1.6.3 Dead-Ending

Appendix Tables that can be used as aids to choose the correct Chicago grips, formed-wire distribution dead ends, and formed-wire service dead ends are

A10–2 Klien Chicago Type Grips for Selected ACSR Conductors
A10.3 Formed Wire Distribution Dead-Ends
A10.4 Formed Wire-Service Dead Ends 6/1 Bare ACSR
A10.5 Dua-Pull Wire-Mesh Grip (Hubbell)

A10.5.2 K-Type Wire-Mesh Grip with Forged Eye
A10–6 K-Type Wire-Mesh with Forged Eye
A10–7 Grips Sizing for Pulling More than One Cable

10.2 Pulling in Underground Cable

10.2.1 Maximum Pulling Tension

Pulling underground cable is all about keeping the pulling tension below the point where the cable and or duct becomes damaged.

Ideally, the design of the duct banks and spacing for the underground cable vaults (UCVs) takes into account the maximum allowable pulling tension for the cable being used. Design should include cable size and weight, the cable's maximum allowable pulling tension, the use of pulling eye or wire-mesh grip, the number of bends in the conduit system, bend radii, the coefficient of friction, and whether cables are pulled dry or lubricated.

The maximum pulling tension for aluminum or copper XLP or EPR cables rated between 600Vv and 35kV, in Table 10–2, are only typical. Consult the manufacturer or engineering for actual maximum allowable pulling tensions.

TABLE 10–2   **Typical Maximum Allowable Pulling Tension for Underground Cable**

Cable Size	1 cable per duct		2 cables per duct		3 cables per duct	
	Grip (lbs.)	Pulling Eye (lbs.)	Grip (lbs.)	Pulling Eye (lbs.)	Grip (lbs.)	Pulling Eye (lbs.)
#4	334	334	668	668	668	668
#2	531	531	1,062	1,062	1,062	1,062
1/0	844	844	1,688	1,688	1,688	1,688
2/0	1,000	1,065	2,000	2,130	2,000	2,130
4/0	1,000	1,693	2,000	3,386	2,000	3,386
250	1,000	2,000	2,000	4,000	2,000	4,000
350	1,000	2,800	2,000	5,600	2,000	5,600
500	1,000	4,000	2,000	8,000	2,000	8,000
750	1,000	6,000	2,000	12,000	2,000	12,000
1,000	1,000	8,000	2,000	16,000	2,000	16,000

10.2.2 Pulling Cable into a Trench

Direct-buried cable can be laid into a trench where it is bedded above and below with stone-free backfill or, sometimes, with special low thermal resistivity backfill, either of which would be brought to the job site. Cable is also available preassembled in a polyethylene conduit (cable-in-conduit system) that can be laid in a trench.

A common stringing method is to pay out the cable from a reel mounted on a truck or trailer that moves along the trench. In this case, while cable is being moved or dropped into the trench cable, it must be protected from abrasion damage.

If pulling a long length into a trench with a winch, it is possible to exceed the allowable pulling tension on a cable. Typically, an adequate number of rollers should be used in such situations to reduce contact with the ground. Rollers should be placed about every 20 feet (6 m), and side rollers should be placed at corners. The pulling tensions given in Table 10–3 are based on a horizontal route with smoothly running supporting and corner rollers.

TABLE 10–3   Pulling Tensions Based on Using Smooth Horizontal Rollers

Route of Trench	% of the weight of the full length of cable
With negligible bends	15–20%
With two bends of 90° each	20–40%
With three bends of 90° each	40–60%

10.2.3 Pulling Cable into a Duct

When the duct is designed to allow cable to be pulled in at less than the maximum allowable tension, it will be assumed that most of the factors in the following list are standard practice and will be carried out.

To make a pull within the tension limits of the cable.

- Immediately prior to the pulling of cable, ensure that the conduit is clean and bell ends smooth.

- Set up the cable reels at the end that is closest to the majority of the cable bends or nearest the sharpest bend so that the cable will not go through these bends until late in the pull.

- Rig the vault at the cable reel end so that a minimum amount of cable bends and a variety of cable guides are made specifically for keeping the cable running smoothly with a minimum amount of tension as it comes off the reel. A quadrant block (see Figure 10–2) allows the cable to make a smooth low-friction 90-degree turn.

- Observe the minimum installation temperature of the cable. If it is necessary to pull cable in very cold weather, preheat the cable immediately before the pull to reduce risk of damage.

- Lubricate (the coefficient of friction in a lubricated duct will be about one-half the coefficient of friction for a dry duct).

- Pull the cable at a constant speed throughout the pull. If the pull is stopped, tension will be increased momentarily when the pull is restarted.

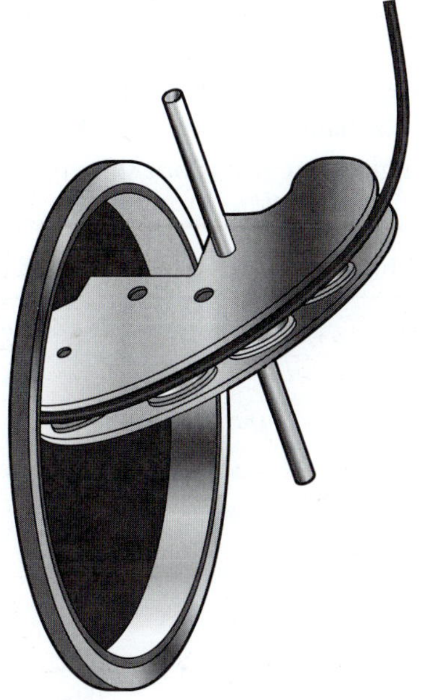

Figure 10–2   A quadrant block.

- A weak link can be installed at the pulling eye to act as a mechanical fuse and to break apart before a cable is overstretched.

10.2.3.1 Maximum Conduit Fill

Conduit fill is the percentage of space, by volume, taken up by cable(s) in the space available in a conduit. Ideally, the conduit specified will be larger than needed to allow for a pull requiring less tension and less risk of jamming.

U.S. National Electric Code specifications for maximum conduit fill:

- One cable = 53%

- Two cables = 31% (Two conductors, same size, form an oval shape.)

- Three cables = 40% (One, three, or more cables tend to form a circular shape.)

To calculate how much area a cable (or cables) will consume:

Cross-sectional area (CSA) of each cable or conduit = $\pi \times r^2$

or

$CSA = D^2 \times 0.7854$
  where D = diameter
  $0.7854 = \pi/4$

Add the CSA for each cable to find the total CSA ÷ conduit CSA × 100 = %fill.

Tables 10–4 and 10–5 show examples of conduit fill in some typical installations and can be used for ballpark estimates. Duct design for higher-voltage installations would be engineered and is not shown.

TABLE 10–4   Conduit Fill for 600V Triplex and Quadruplex

| Size and Type | | | Conduit Diameter | | |
of Cable	*2 inches*	*3 inches*	*4 inches*	*5 inches*	*6 inches*
1/0 triplex	28%	13%	—	—	—
4/0 triplex	48%	22%	—	—	—
350 triplex	—	34%	20%	—	—
1000 triplex	—	—	49%	32%	—
1/0 quadruplex	35%	16%	—	—	—
4/0 quadruplex	—	27%	16%	—	—
350 quadruplex	—	42%	25%	16%	—
1000 quadruplex	—	—	61%	40%	28%

TABLE 10–5   Conduit Fill for 25kV Cable

| Size of Cable | Conduit Diameter | | |
	4 inches	*5 inches*	*6 inches*
1/0	41%	—	—
600 kcmil	57%	37%	26%
1,100 kcmil	—	48%	34%

10.2.3.2 Avoiding Cable Jamming in Duct

When three cables (often secondary) are pulled through conduit, they can jam when the triangular configuration is lost and the cables roll over and line up in parallel as they are pulled around a bend. The cables jam when the combined outside diameters of the three cables equal the inside diameter of the conduit. An unfavorable ratio of cable outside diameter (OD) to conduit inside diameter (ID) is called a *jam ratio.*

$J = 1.05\ ID \div OD$

where J = jam ratio

           ID = inside diameter of conduit (inches or mm)

           OD = outside diameter of the cable (inches or mm)

The additional 1.05 factor may be caused by the ovalization of the conduit at the bend.

Table 10–6 indicates that there is some probability of three cables jamming from 29% to 52% conduit fill. If calculations show that jamming is likely, it is not automatic. A conduit with two bends may reduce the chance of the individual cables lining up. The probability of jamming is minimized with lower pulling tension, which may be achieved with lubrication.

TABLE 10–6   Probability of Jamming

Jam Ratio	Probability of Jamming
2.4–2.5	Low
2.5–2.6	Medium
2.6–2.9	High
2.9–3.0	Medium
3.0–3.2	Low

10.2.3.3 Pulling with a Wire-Mesh Grip

While a manufacturer- or field-installed pulling eye is the best way to pull a cable, a wire-mesh grip is often used. Table A10–6 in this chapter's appendix shows one type of many that are available from different manufacturers. It may take some research to ensure the correct strength for the size of conductor or rope chosen. Table A10–7 in this chapter's appendix has tables showing the sizes needed for multiple cables.

Formed Tie Wires

This table identifies the ties for some of the more common conductors. Read the identification tag to confirm the correct size and type of conductor, type of insulator, neck size of insulator, top tie, side tie, double tie.

A10.1 Hubbell Super Top Tie

TABLE A10–1   Hubbell Super Top Tie

	STT60	STT70	STT80	STT90	STT100	STT110	STT120	SST130
	Yellow	Blue	Black	Pink	Green	Brown	Violet	Gold
Minimum Range	0.361"	0.410"	0.461"	0.517"	0.585"	0.665"	0.756"	0.860"
	9.2mm	10.4mm	11.7mm	13.1mm	14.9mm	16.9mm	19.2 mm	21.8mm
Maximum Range	409"	0.460"	0.516"	0.584"	0.664"	0.755"	0.859"	0.977"
	10.4mm	11.7mm	13.1mm	14.8mm	16.9 mm	19.2mm	21.8 mm	24.8mm
AAC	1/0	2/0	3/0	4/0	266.8	336.4	477	600
				250		350	556.5	715.5
ACSR	1/0	2/0		4/0	266.8	336.4	397.5	556.5
						397.5	477.0	636l
AAAC	121.5	156.1	196.8	248.2	307.1	394.6	466.3	652.8
Insulator				Insulator Neck Diameter 2.25" (57.1 mm) – 3.5" (88.9 mm)				

A10.2 Dead-Ending Hardware

TABLE A10–2   Klien Chicago Type Grips for Selected ACSR Conductors

Tables are also available for bare copper, all aluminum, compact aluminum, zinc coated steel cable, polyethylene, alumoweld and other types of conductor.

Klien #	WLL lbs.	WLL kN	AWG/kcm il	mm²	Stranding	OD in.
1656-20	750	2.5	4	21.2	6/1	0.250
1656-20	1,000	3.7	2	33.6	6/1	0.316
1656-30	2,800	12.3	2	33.6	4/3	0.399
1656-30	2,000	8.8	1/0	53.5	6/1	0.398
1656-30	2,000	8.8	2/0	67.5	6/1	0.447
1656-30	2,000	8.8	3/0	85	6/1	0.563
1656-40	3,000	13.3	4/0	107	6/1	0.447
1656-40	4,200	18.6	336.4	170	26/7	0.721
1656-40	2,700	12.0	336.4	170	18/1	0.685
1656-40	4,000	17.7	336.4	170	30/7	0.741
1628-30	7,000	31.1	477	242	26/7	0.858
1628-30	7,000	31.1	477	242	30/7	0.883
1628-30	7,350	32.7	500	253	30/7	0.904
1628-30	4,200	18.6	500	253	54/7	0.866
1628-30	6700	29.8	556.5	282	26/7	0.927
1628-30	8,000	35.5	605	307	54/7	0.953
1628-30	7,400	32.9	666.6	338	54/7	1.0
1628-30	11,000	49.0	795	403	26/7	1.108
1628-30	11,500	51.1	795	403	30/19	1.140
1628-30	8,700	38.7	795	403	54/7	1.093

A10.3 Formed Wire Distribution Dead-Ends

Manufacturers' tables must be consulted for other types of conductor or jacketed conductor.

TABLE A10–3   Formed Wire-Distribution Dead Ends

ACSR	All Aluminum	Aluminum Alloy	Compact ACSR	Color Code
#6 6/1	#6 7W	#6 7W	#6 6/1	blue
#4 6/1	#4 7W	#4 7W	#4 6/1	orange
#2 6/1	#2 7W	#2 7W	#2 6/1	red
#1 6/1	#1 7W	#1 7W	#1 6/1	green
1/0 6/1	1/0 7W	1/0 7W	1/0 6/1	yellow
2/0 6/1	2/0 7W	2/0 7W	2/0 6/1	blue

(continued)

TABLE A10–3   *(continued)*

ACSR	All Aluminum	Aluminum Alloy	Compact ACSR	Color Code
3/0 6/1	3/0 7W	3/0 7W	3/0 6/1	orange
4/0 6/1	4/0 7W	4/0 7W	4/0 6/1	red
266.8, 18/1	266.8 19W	266.8 19W	336.4 18/1	black
336.4, 18/1	336.4 19W	336.4 19W	397.5 18/1	green
397.5, 18/1	450, 19W	397.5, 19W	477, 18/1	orange
477, 36/1	477, 19W		556, 19W	orange
477, 18/1	500, 37W			orange
556.5, 36/1	556.5, 37W	477, 19W	636, 18/1	blue
605, 36/1	636, 37W		795, 19W	blue
636, 18/1	650, 61W			blue
666.6, 36/1	715.5, 37W	636, 37W	874.5, 37W	brown
715.5, 36/1	750, 61W		954, 37W	brown
795, 36/1	795, 61W			brown
874.5, 36/1	874.5, 61W	795, 37W		orange
954, 36/1	954, 61W			orange
1033.5, 36/1	1033.5, 61W			orange

A10.4 Formed Wire-Service Dead Ends 6/1 Bare ACSR

TABLE A10–4   **Formed Wire-Distribution Dead Ends 6/1 Bare ACSR**

Nominal Conductor Size	Conductor Diameter	Color Code
#6	00.169 to 00.198	blue
#5	0.199 to 0.224	white
#4	0.225 to 0.257	orange
#3	0.258 to 0.289	black
#2	0.290 to 0.325	red
#1	0.326 to 0.360	green
1/0	0.361 to 0.400	yellow
2/0	0.401 to0.450	blue
3/0	0.451 to 0.510	orange
4/0	0.511 to 0.580	red

A10.5 Dua-Pull Wire-Mesh Grips (Hubbell)

Wire-mesh grips are found in many types and from many different manufacturers. It may take some research to ensure the correct strength for the size of conductor or rope chosen.

A10.5.1 Kellems Dua Pull Grips

These grips are suitable for ACSR, ACAR, all aluminum and copper conductor, ground wires, messenger strands, and synthetic ropes.

TABLE A10–5   Kellems Dua Pull Grips

Color	Conductor Diameter Range (in.)	(mm)	Rope Diameter Range (in.)	(mm)	Breaking Strength (approx.) (lbs.)	(N)
Black	.19–.37	5–9	.25–.65	6–17	6,500	28,912
Green	.38–.62	10–16	.50–.90	13–23	14,000	62,272
Red	.63–.87	16–22	.75–1.10	19–28	20,000	88,960
Blue	.88–1.12	22–28	1.00–1.50	25–38	30,600	136,109
Yellow	1.13–1.37	29–35	1.25–1.70	32–43	46,800	208,166
Aluminum	1.38–1.90	35–48	1.50–2.10	38–53	66,500	295,792

A10.5.2 K-Type Wire-Mesh Grip with Forged Eye

This type of grip is designed for pulling underground cables. The forged eyes are designed to mate with a swivel or shackle.

TABLE A10–6   K-Type Wire-Mesh Grip with Forged Eye

Description	Cable Diameter Range (in.)	(mm)	Breaking Strength (approx.) (lbs.)	(N)
033-01-011	.50–.61	13–15	5,600	24,909
033-01-012	.62–.74	16–19	6,800	30,246
033-01-024	.75–.99	19–25	9,600	42,701
033-01-025	1.00–1.49	25–38	16,400	72,947
033-01-026	1.50–1.99	38–51	16,400	72,947
033-01-027	2.00–2.49	51-63	27,200	120,986
033-01-028	2.50–2.99	64-76	33,000	146,784
033-01-029	3.00–3.49	76-89	41,000	182,368
033-01-030	3.50–3.99	89-101	48,000	213,504
033-01-031	4.00–4.49	102-114	48,000	213,504
033-01-039	4.50–4.99	114-127	48,000	213,504
033-01-047	5.00–5.99	127–152	48,000	213,504
033-01-045	6.00–6.99	152–178	48,000	213,504

A10.5.3 Grips Sizing for Pulling More than One Cable

TABLE A10–7   Grips Sizing for Pulling More than One Cable

Individual Cable Outside Diameters, inches (mm)					Grip Diameter Range in. (mm)
2	3	4	5	6 & 7	
.30–.38 (8–10)	.25–.31 (6–8)	.22–.27 (6–7)	.19–.24 (5–6)	.17–.22 (4–6)	.50–.61 (13–15)
.38–.44 (10–11)	.31–.36 (8–9)	.27–.31 (7–8)	.24–.29 (6–7)	.22–.26 (6–7)	.62–.74 (16–19)
.44–.59 (11–15)	.36–.49 (9–12)	.31–.42 (8–11)	.29–.38 (7–10)	26–.34 (7–9)	.75–.99 (19–25)
.59–.75 (15–19)	.49–.63 (12–16)	.42–.54 (11–14)	.38–.48 (10–12)	.34–.43 (9–11)	1.00–1.24 (25–31)
.75–.90 (19–23)	.63–.76 (16–19)	.54–.65 (14–17)	.48–.58 (12–15)	.43–.52 (11–13)	1.25–1.49 (31–38)
.90–1.07 (23–27)	.76–.89 (19–23)	.65–.77 (17–19)	.58–.67 (15–17)	.52–.60 (13–15)	1.50–1.74 (38–44)
10.7–1.22 (27–31)	.89–1.02 (23–26)	.77–.88 (19–20)	.67–.77 (17–19)	.60–.69 (15–18)	1.75–1.99 (44–51)
1.22–1.53 (31–39)	1.02–1.28 (26–33)	.88–1.10 (20–28)	.77–.96 (19–24)	.69–.86 (18–22)	2.00–2.49 (51–63)
1.53–1.83 (39–46)	1.28–1.53 (33–39)	1.10–1.32 (28–34)	.96–1.16 (24–29)	.86–1.03 (22–26)	2.50–2.99 (63–76)
1.83–2.14 (46–54)	1.53–1.79 (39–45)	1.32–1.54 (34–39)	1.16–1.35 (29–34)	1.03–1.20 (26–30)	3.00–3.49 (76–89)
2.14–2.44 (54–62)	1.79–2.05 (45–52)	1.54–1.76 (39–45)	1.35–1.54 (34–39)	1.20–1.37 (30–35)	3.50–3.99 (89–101)
2.44–2.75 (62–70)	2.05–2.30 (52–58)	1.76–1.98 (45–50)	1.54–1.74 (39–44)	1.37–1.55 (35–39)	4.00–4.49 (101–114)
2.75–3.06 (70–78)	2.30–2.56 (58–65)	1.98–2.20 (50–56)	1.74–1.93 (44–49)	1.55–1.72 (39–44)	4.50–4.99 (114–127)

CHAPTER 11

Overhead Connections and Splices

11.1 Identify the Conductor

11.1.1 Choose the Correct Sleeve or Connector

Before a connector or sleeve can be chosen, the type and size of conductor must be identified. Only the conductive material is used when discussing conductor size, so the type of stranding and type and size of steel wires in a conductor will change the diameter. For example, at least three 336.4-circular-mil ACSR conductors with three different diameters are available.

Section 11.1.2 and the appendix at the end of this chapter may help identify the type of conductor to be spliced or connected. In part, the appendix shows the weight of conductors because that is needed to calculate the weight to be lifted, the tension, and the bisect tension. Tensile strength and ampacity are only needed by design engineers, so that information is not included.

11.1.2 Some Conductor Types

Conductor Type	Identifying Aids
Aluminum conductor, steel-reinforced (ACSR)	The reinforcing wires may be in a central core or distributed throughout the cable. A galvanized or aluminized coating will be on the steel wires. A large variety of stranding arrangements are available. For example, a ratio of aluminum strands to steel strands can be of 6/1, 18/1, 20/7, 24/7, 26/7, or 30/7.
All-aluminum conductor (AAC), also called aluminum-stranded conductor (ASC)	Used in distribution in which shorter spans do not require steel reinforcement for strength. Corrosion resistance of aluminum has made AAC a conductor of choice in coastal areas.
All-aluminum-alloy conductor (AAAC)	Looks like AAC, but it is a conductor with higher-strength and individual aluminum-alloy strands. The individual strands are much stiffer and thus more difficult to bend than the aluminum in AAC.
Aluminum conductor, aluminum-alloy-reinforced (ACAR)	Looks like AAC and AAAC, but ACAR is a combination of aluminum and aluminum alloy strands and will have easy-to-bend stands and hard-to-bend strands.
Aluminum-alloy conductor, steel-reinforced (AACSR)	AACSR is an aluminum-alloy conductor (AAAC) with steel strands added for more strength. It is a very high-strength conductor used for extra-long spans or as a messenger/neutral cable for spun secondary.

(continued)

11.1.2 Some Conductor Types *(continued)*

Conductor Type	Identifying Aids
ACSR with trapezoid-shaped aluminum wires (ACSR/TW)	ACSR/TW is a compact conductor with more aluminum area in a smaller diameter. For a given diameter, this provides an increase in the current-carrying capacity using less sag and longer spans.
Aluminum composite conductor reinforced (ACCR)	Similar to ACSR except that the core is a composite material instead of steel. The core consists of a number of individual composite wires that look like the steel wires in ACSR, but each composite wire provides strength at about half the weight of steel. An existing line can be restrung to approximately double its ampacity with no increase in conductor diameter, less weight, and no requirement to upgrade the structures.
Aluminum conductor composite core, wrapped with trapezoid-shaped aluminum wires (ACCC/TW)	A variation of ACCR with trapezoid-shaped aluminum wires similar to those used in ACSR/TW conductor.
ACSR self-dampening conductor	Constructed of steel wires in the core surrounded by layers of trapezoid-shaped aluminum wires. It is designed to keep a small gap between the layers of wire while under tension. The interaction of the different natural vibration frequencies of the steel core and aluminum layers provides an internal dampening effect.
Vibration-resistant (VR) or twisted (T-2) conductor	Composed of two identical conductors twisted together, giving the conductor a spiraling figure-eight shape. The spiraled shape presents a continuously changing conductor diameter to the wind, which disrupts the force of the wind on the conductor. During repair or splicing, each of these two conductors must have its own grips and hoist installed. Any splices should be about 15 ft. (3 m) apart. *Oval conductor* is another vibration-resistant conductor that also provides continuously changing diameters to the wind.

(continued)

11.1.2 Some Conductor Types *(continued)*

Conductor Type	Identifying Aids
Aluminum-weld (AW) (also called aluminum-clad) conductor	Has aluminum cladding bonded onto each individual strand of high-strength steel wire. A wire size of "7 No. 5" means there are 7 strands of number-5 wire. Similarly copper-weld (CW) (also called copper-clad) conductor, has copper cladding bonded to steel wires.
Spacer cable/Tree wire	Can be AAC or ACSR conductor covered with cross-linked polyethylene (XLPE) or high molecular-weight low-density (HMW-LDPE) or high-density polyethylene (HMW-HDPE). The jacket is not considered insulation but a cover that prevents an immediate short circuit when the conductor makes contact with trees. The cable must be treated as a bare conductor for working purposes because the cover may have deteriorated without the deterioration being readily visible.

11.2 Making Connections

11.2.1 Hazards when Making Connections

If	Then
Cutting or joining a live conductor	Anytime a connection to a live circuit must be made or broken, a jumper or lead first must be installed with a shotgun stick. Taking a shortcut by using rubber gloves or working barehand to make or break a connection causes serious accidents. All too frequently, unforeseen situations have occurred, such as breaking too much load, connecting into a short circuit, a connector failure, a worker's body accidently making contact between the two elements, or a high-charging current energizing a line or transformer.
Cutting or joining an isolated and grounded conductor	Install a jumper. A current flow in a grounded conductor is very likely to be induced from neighboring circuits. The resistance across the open point is infinite; therefore, the voltage drop (IR) is at full-line voltage. This

(continued)

11.2.1 Hazards when Making Connections *(continued)*

If	Then
	full-line voltage (recovery voltage) appears across the two ends of the cut conductor.
Installing a live-line clamp	Install on a stirrup (bale) and not directly on the conductor. A poor connection directly on the conductor can cause the connector to heat up and the conductor to burn apart.
Working with small conductor	• A live primary conductor, such as a brittle #6 copper or #4 ACSR with the steel core rusted away, can break with very little disturbance. • Do not do any live-line work on this conductor. In the past it has broken from the shock of firing on a wedge connector, from using a hot stick to knock ice off the conductor, and while removing a live-line clamp. • Do not stand under this conductor when any work is being done on it.

11.2.2 Making Quality Connections

If	Then
Making aluminum connections	High-resistance oxides form very quickly on aluminum. Aluminum oxide is a highly resistant transparent film that forms immediately on the surface of aluminum when exposed to air. Even aluminum conductor that looks clean and bright must be cleaned. Use oxidation-inhibiting joint compounds to prevent reoxidation in the connections. Many old bolted connectors, such as parallel groove clamps and split bolts that are good for copper, will eventually fail on aluminum.
Making copper connections	Copper oxide, which is the green coating on the wire, is somewhat conductive, and that is why copper connections rarely burn off.
Making an aluminum to copper connection	Approved connectors that join aluminum and copper have a divider (often with a cadmium surface) between the two metals. The copper should be on the

(continued)

11.2.2 Making Quality Connections *(continued)*

If	Then
	bottom to prevent the copper oxide from leaching over and corroding the aluminum.
Connecting pads and lugs	These are cleaned and then bolted together. Belleville (spring) washers are used to keep constant pressure on the connection by compensating for the expansion and contraction of the bolts and connection. They must be properly torqued so that they are compressed to only 80% of their height; if no bevel is left in the washer, the connection is too tight and expansion could cause a failure.
Making a connection with a fired-on-wedge connector	1. Only trained personnel makes the connection. 2. The tool is maintained. 3. The daily check indicates that the fail safe is working. 4. Choose the correct-size tool (gun), cartridge, and connector (see Appendix 11).
Installing a compression connector	1. Choose the correct connector and die. 2. To prevent a possible musculoskeletal injury, use a hydraulic- or battery-operated press, not a mechanical press.
Installing a Versa-Crimp connector	One size connector fits many. The dieless Versa-Crimp compression tool operates on a fixed pressure until the preset hydraulic relief valve in the tool senses that the proper force has been applied to the crimp. The valve then "pops off" or releases the pressure in the crimp tool.
Connecting using exothermic or thermit welding (Cadweld, Thermoweld, Techweld)	The connection is made by bonding two materials by melting them together. An electrical connection is made by pouring superheated, molten-copper alloy into a mold that contains the conductors being joined. The heating is due to a chemical reaction in which temperatures reach 4,000°F (2,200°C). The connector joins together materials such as copper, steel (plain or galvanized), clad steel, bronze/brass, and stainless steel. **Note:** A wet mold can lead to an explosion, and hot molds can be fire hazards.

11.3 Making a Splice

11.3.1 Procedure for Installing a Two-Piece Compression Sleeve

Step	Action	Details
1	Choose the correct sleeves and dies.	The material used to splice a large ACSR conductor consists of a steel sleeve for strength and an aluminum splice body to carry the electrical current. Refer to company documents or manufacturers' catalogs to ensure that the correct size and type of sleeves are chosen for the conductor size and type. Each steel and aluminum sleeve is stamped with the die size and usually the conductor type and size, along with information related to the manufacturer. It is important to use the dies specified on the sleeve because no standard is interchangeable among manufacturers.
2	Slide aluminum sleeve onto one end of the conductor.	Slide the aluminum splice body over one end of the conductor far enough to allow the installation of the steel sleeve. This is the most embarrassing step to forget, especially when working hot on a tensioned conductor and not noticing the absence of the aluminum splice body until after the steel sleeve is pressed. A piece of conductor and two splices are required to fix the situation.
3	Expose the steel core.	The aluminum strands on each end of the conductor being spliced are cut back, square with the conductor, half the length of the steel sleeve plus at least another 1/2 in. (1 cm) because the steel sleeve will expand lengthwise as it is compressed. Ideally, a cable trimmer with a cable trimmer bushing is used to make a square cut. If using a hacksaw, it is critical *not* to nick the steel wires. Taping the aluminum where the cut is to be made will prevent the individual strands from bending while being cut. After the steel wire is exposed, make or keep it straight and clean it. A small piece of wire wrapped near the end of the steel will prevent the strands from unraveling.

(continued)

11.3.1 Procedure for Installing a Two-Piece Compression Sleeve *(continued)*

Step	Action	Details
4	Insert the steel core into the steel sleeve.	The steel must be measured and marked, often with tape to indicate when the steel core reaches the middle of the sleeve. If the steel sleeve has a crimp in the center indicating the middle, it will be fairly obvious when the center is reached. The rated strength of the splice will not be achieved if the steel ends are not in the center. The wire wrapped around the end of the steel stranding should slide back as the steel is inserted into the sleeve.
5	Press the steel sleeve.	Use one of the many suitable compression tools and the die size and type that are stamped on the sleeve. Make the first compression directly over the center, capturing both ends of the steel core. Press from the center toward each end and overlap each compression by about 10%. Depending on the type of die and other factors, the sharp edges may have to be filed off the completed sleeve and the sleeve may need straightening. If the sleeve is being installed on tensioned conductor, back off the hoist and remove the grips so that you have more room to work with the aluminum and birdcaging will be less likely when pressing the aluminum.
6	Clean the aluminum stranding.	If the conductor is very black, it can be cleaned using a caustic soda such as lye. Before the steel sleeve is installed, each end of the wire should be inserted into hot water and lye. Eye protection must be worn. Ideally, a special "lye pot" made from a heavy-gauge steel is used for this method of cleaning conductor. The conductor should be rinsed after the cleaning. The standard method of cleaning is to unravel (without bending) the aluminum strands and to wipe each strand with an abrasive cleaning pad or fine steel wool coated with an oxide inhibitor. This can be very dirty work, especially when working hot

(continued)

11.3.1 Procedure for Installing a Two-Piece Compression Sleeve *(continued)*

Step	Action	Details
		barehand or with rubber gloves. Remember that the no-oxide inhibitor is conductive.
		It is not normally necessary to unravel the last layer of aluminum strands next to the steel core.
7	Slide the aluminum body over the sleeve.	Wrap the aluminum strands back into place so that it can be inserted into the sleeve.
		Measure and mark or tape the aluminum conductor so that when the aluminum body is slid into place it is directly over the center of the steel sleeve.
8	Insert the joint into the filler hole.	One or two filler holes are typically in the aluminum body of a two-piece sleeve so that the cavity around the steel sleeve can be filled with joint compound. Joint compound is conductive and helps prevent oxidation. Joint compound is added with a grease gun or caulking gun to supply pressure. The compound is added until it begins to flow out between the aluminum body and the conductor. An aluminum plug or pin is then hammered into the filler hole until it is flush with the sleeve body.
		Use one of the many suitable compression tools and the die size and type that are stamped on the splice body. The starting point and direction are important for preventing high-stress points in the joint.
9	Press the aluminum sleeve.	Make the first compressions on each side of the steel sleeve and move out toward the ends, overlapping each compression. Make no compressions over the top of the steel sleeve. The sleeve will expand out as the compressions are made. Remove any tape that marked the center.
		Pressing a one-piece sleeve is done in similar fashion except that the press can start on either side of the middle and can press full length each way. A terminal is pressed by starting at a point closest to the pad (tongue).
		The sharpe edges of joints, especially on high-voltage lines, must be filed off to make the joint smooth.

11.3.2 Procedure for Installing an Automatic Splice

Step	Action	Details
1	Select the proper splice	Check the conductor type and size stamped on the shell of the sleeve. Automatic splices can typically be used on ACSR, ASC, and AAAC. The end caps are color coded:

Conductor Size	End Cap Colour
4 ACSR	orange
2 ACSR/AASC	red
1/0 ACSR/AASC	yellow
2/0 ACSR/AASC	grey
3/0 ACSR/AASC	black
4/0 ACSR/AASC	pink
336.4 ASC	green
4 to #2 ACSR/AASC	Reducer orange/red
1/0 to 2/0 ACSR/AASC	Reducer yellow/grey

Fargo has range-taking automatic splices that are used for more than one size—for example, a sleeve for 3/0 or 4/0 has black bands over pink funnel guides.

Step	Action	Details
2	Cut the conductor.	Straighten the conductor to remove any coil or curvature. Ensure the conductor is straight and free of burrs or damage. On existing conductor, put the conductor grips as far apart as possible to allow the conductor ends to be inserted in the splice without undue conductor bending.
3	Prepare the conductor to insert into the sleeve.	Clean the conductor; do not apply a grit type inhibitor. Measure and tape the conductor so that it will show when the conductor is fully inserted.
4	Insert the conductor to the center mark.	Insert the conductor into the pilot cup and push the conductor into the automatic splice in a single smooth motion until it strikes the center stop. If the conductor does not go in smoothly, conductor strands are probably getting caught between the jaws inside the sleeve. Forcing and twisting the conductor will not work. Remove the sleeve and start again with a new one.

(continued)

11.3.2 **Procedure for Installing an Automatic Splice** *(continued)*

Step	Action	Details
5	Ensure the sleeve jaws are set.	Set the jaws by giving the splice a sharp pull. The jaws must be set before allowing line tension. Do not tap the sleeve to set the jaws. As tension is applied, the conductor may pull back a small amount (¼ to ½ inch).
Notes		• Automatic splices are not to be used in slack spans or jumpers or for stringing. • Automatic splices shall not be reused. • A utility may restrict use of automatic splices in coastal areas because of corrosion concerns, especially if it also does not use ACSR because the steel core corrodes.

11.3.3 Implosive Sleeves and Terminals

Implosive sleeves are ideal for major transmission projects. The finished product is void free, totally compressed, smooth, straight, and has no birdcaging. After the usual conductor cleaning and preparations (a separate steel sleeve is used on ACSR), the implosive (not explosive) charge is set off and crushes the sleeve and conductor into a solid piece of metal. The sleeves can be used during stringing and will go through stringing blocks.

The actual implosion releases a lot of energy and is very loud. People living nearby are usually alerted to the time the implosion is to take place. It is therefore typical to have all the necessary joints prepared and imploded at one time. On one side of a dead-end tower with three bundled conductors of four each, twelve implosions may be set off simultaneously.

Conductor Data

A11.1 Bare Aluminum Conductor Steel Reinforced (ACSR)

TABLE A11–1   Bare Aluminum Conductor Steel Reinforced (ACSR)

AWG/kcmil	mm²	Stranding	OD (in.)	Weight (lbs./ 1,000 ft.)	Code Word
4	21.2	6/1	.250	57.4	Swan
2	33.6	6/1	.316	91.3	Sparrow
1/0	53.5	6/1	.398	145.3	Raven
2/0	67.5	6/1	.447	183.1	Quail
3/0	85	6/1	.563	230.8	Pigeon
4/0	107	6/1	.447	291.1	Penguin
266.8	135	18/1	.609	289.5	Waxwing
266.8	135	26/7	.642	289.5	Partridge
336.4	170	18/1	.684	365.3	Merlin
336.4	170	26/7	.721	462.6	Linnet
477	242	26/7	.858	656.0	Hawk
477	242	30/7	.883	747.4	Hen
556.5	282	24/7	.914	716.8	Parakeet
556.5	282	26/7	.927	766.0	Dove
636	322	26/7	.990	875.2	Grosbeak

(continued)

TABLE A11–1   *(continued)*

AWG/kcmil	mm²	Stranding	OD (in.)	Weight (lbs./ 1,000 ft.)	Code Word
666.6	338	24/7	1.000	858.9	Flamingo
795	403	26/7	1.108	1094.0	Drake
954	483	54/7	1.196	1229.0	Cardinal
1192.5	604	45/7	1.302	1344.0	Bunting

A11.2 Bare All Aluminum Conductor (AAC)

TABLE A11–2   Bare All Aluminum Conductor (AAC)

Size	Stranding	OD (in.)	Weight (lbs./1,000 ft.)	Code Word
1/0	7	.368	99.1	Poppy
1/0	19	.371	99.0	Geranium
2/0	7	.416	124.9	Aster
2/0	19	.416	125.0	Buttercup
3/0	7	.464	157.5	Phlox
3/0	19	.467	158.0	Primrose
4/0	7	.522	198.7	Oxlip
4/0	19	.525	198.0	Sunflower
250,000	19	.574	234.6	Valerian
250,000	37	.575	235.0	Dandelion
266,800	7	.586	250.6	Daisy
266,800	19	.593	250.4	Laurel
336,400	19	.666	316.0	Tulip
350,000	19	.679	328.4	Daffodil
350,000	37	.681	329.0	Gardenia
397,500	19	.724	373.4	Canna
477,000	19	.793	447.5	Cosmos
477,000	37	.795	447.4	Syringa
500,000	19	.811	469.2	Zinnia
500,000	37	.813	469.0	Hyacinth

(continued)

TABLE A11–2 *(continued)*

Size	Stranding	OD (in.)	Weight (lbs./1,000 ft.)	Code Word
556,500	19	.856	522.1	Dahlia
556,500	37	.858	522.0	Mistletoe
600,000	61	.893	564.0	Lotus
636,000	37	.918	596.4	Orchid
700,000	61	.964	656.8	Flag
715,500	37	.974	672.0	Violet
750,000	61	.998	704.3	Cattail
795,000	37	1.026	746.4	Arbutus
795,000	61	1.028	746.7	Lilac
954,000	37	1.124	895.8	Magnolia
954,000	61	1.126	896.1	Goldenrod
1,000,000	61	1.152	938.2	Camellia
1,033,500	37	1.170	970.0	Bluebell
1,033,500	61	1.172	970.6	Larkspur

A11.3 Bare ACSR/AW with Alumaclad Core

ACSR/AW is primarily used as the phase conductor for overhead transmission and distribution. It is preferred over conventional ACSR because of its higher corrosion resistance, lighter weight, longer service life, and reduced power loss.

TABLE A11–3   Bare ACSR/AW with Alumaclad Core

Size	OD (in.)	Stranding	Weight (lbs./1,000 ft.)	Code Word
2	0.316	6/1	86.8	Sparrow/AW
1/0	0.398	6/1	138.2	Raven/AW
2/0	0.447	6/1	174.2	Quail/AW
3/0	0.502	6/1	219.4	Pigeon/AW
4/0	0.563	6/1	276.8	Penguin/AW
266.8	0.609	18/1	283.5	Waxwing/AW
266.8	0.642	26/7	349.6	Partridge/AW
336.4	0.684	18/1	357.7	Merlin/AW
336.4	0.721	26/7	441.1	Linnet/AW
336.4	0.741	30/7	495.1	Oriole/AW

TABLE A11–3 *(continued)*

Size	OD (in.)	Stranding	Weight (lbs./1,000 ft.)	Code Word
477.0	0.814	18/1	507.2	Pelican/AW
477.0	0.846	24/7	589.4	Flicker/AW
477.0	0.858	26/7	625.4	Hawk/AW
477.0	0.883	30/7	702.0	Hen/AW
556.5	0.879	18/1	591.5	Osprey/AW
556.5	0.914	24/7	687.5	Parakeet/AW
556.5	0.927	26/7	729.1	Dove/AW
556.5	0.953	30/7	819.0	Eagle/AW
636.0	0.940	18/1	676.5	Kingbird/AW
636.0	0.977	24/7	785.6	Rook/AW
636.0	0.990	26/7	833.0	Grosbeak/AW
636.0	1.019	30/19	928.9	Egret/AW
666.6	1.000	24/7	823.7	Flamingo/AW
666.6	1.014	26/7	873.0	Gannet/AW
795.0	1.063	45/7	873.4	Tern/AW
795.0	1.092	24/7	981.8	Cuckoo/AW
795.0	1.093	54/7	982.8	Condor/AW
795.0	1.108	26/7	1042.0	Drake/AW
795.0	1.140	30/19	1161.0	Mallard/AW
954.0	1.140	36/1	955.0	Catbird/AW
954.0	1.165	45/7	1049.0	Rail/AW
954.0	1.196	54/7	1178.0	Cardinal/AW
1192.5	1.302	45/7	1311.0	Bunting/AW
1272.0	1.345	45/7	1398.0	Bittern/AW
1272.0	1.382	54/19	1570.0	Pheasant/AW

A11.4 Bare Copper

TABLE A11–4   Bare Copper

Size	Stranding	OD (in.)	Weight (lbs.)/ 1,000 ft.
4	solid	0.204	126.3
2	7 HD	0.292	204.9
1/0	7 MHD	0.368	325.8
2/0	7 HD	0.414	410.8
2/0	19 MHD	0.419	410.9
3/0	7 HD	0.464	518.1
4/0	12 HD	0.552	650.2
4/0	19 HD	0.529	653.3
450	19 HD	0.770	1,389
500	37 MHD	0.813	1,544
1000	61 MHD	1.152	3,088
1250	61 MHD	1.288	3,859
1500	61 HD	1.411	4,631

HD = hard drawn

MHD = medium hard drawn

A11.5 Bare Alumaclad (Alumoweld) Conductor

These stranded conductors of aluminum-clad steel wires are commonly used for overhead shield wire. **Note:** The diameter and circular mils (mm²) of copperweld are the same as for alumoweld with equivalent standing. Copperweld is approximately 25% heavier than alulmoweld.

TABLE A11–5   Bare Alumaclad (Alumoweld) Conductor

Size Designation	# of strands	AWG of Strands	Diameter (in.)	Diameter (mm)	Lbs./ 1,000 ft.	kg/km	(kc mils)	mm²
37 # 6	37	#6	1.13	28.8	2,222	3,307	971.3	492.2
37 # 8	37	#8	0.899	22.9	1,398	2,080	610.9	309.5
37 # 10	37	#10	0.713	17.9	879.0	1,308	384.2	194.7
19 # 6	19	#6	0.810	20.6	1,134	1,688	498.8	252.7
19 # 8	19	#8	0.642	16.3	713.5	1,062	313.7	158.9

(continued)

TABLE A11–5 *(continued)*

Size Designation	# of strands	AWG of Strands	Diameter (in.)	Diameter (mm)	Lbs./ 1,000 ft.	kg/km	(kc mils)	mm²
19 # 10	19	#10	0.509	12.9	448.7	666.7	197.3	99.9
7 # 6	7	#6	0.486	12.4	416.3	619.5	183.8	93.1
7 # 8	7	#8	0.385	9.8	261.8	389.6	115.6	58.6
7 # 10	7	#10	0.306	7.8	164.7	245.1	72.68	36.8
7 # 12	7	#12	0.242	6.2	103.6	154.2	45.71	23.2
3 # 6	3	#6	0.349	8.9	178.1	265.0	78.75	39.9
3 # 8	3	#8	0.277	7.0	112.0	166.7	49.53	25.1
3 # 10	3	#10	0.220	5.6	70.43	104.8	31,150	15.8

A11.6 15kV AAC Spacer Cable–Tree Wire

Tree cable is an uninsulated conductor. The covering is intended to prevent shorts or flashovers with other objects. Conductors are covered with XLPE or with HMW low-density or high-density polyethylene.

TABLE A11–6   15kV AAC Spacer Cable–Tree Wire

Size	Stranding	OD (in.)	Weight (lbs./1,000 ft.)		
			XLPE	*HMW Poly*	*HD Poly*
2	7	.583	161.0	144.7	147.3
1/0	19	.662	214.4	196.5	199.6
2/0	19	.706	250.5	230.7	234.1
3/0	19	.756	295.1	272.8	276.5
4/0	19	.812	349.8	324.5	328.6
250	37	.858	392.3	369.4	373.7
266.8	19	.874	416.4	388.2	392.7
336.4	37	.947	493.5	467.4	472.3
350	37	.961	510.1	482.9	487.8
400	37	1.006	567.1	538.3	543.6
450	37	1.049	624.1	593.4	598.9
477	37	1.071	654.7	623.0	628.6
500	37	1.089	680.4	648.0	653.8

A11.7 15kV ACSR Spacer Cable–Tree Wire

Tree cable is an uninsulated conductor. The covering is intended to prevent shorts or flashovers with other objects. Conductors are covered with XLPE, HMW low-density or high-density polyethylene.

TABLE A11–7   15kV ACSR Spacer Cable–Tree Wire

Size	Stranding	OD (in.)	XLPE	HMW Poly	HD Poly
			\(Weight (lbs./1,000 ft.)\)		
2	6/1	.616	200.2	186.0	189.0
1/0	6/1	.698	276.0	258.9	262.6
2/0	6/1	.747	327.6	308.7	312.7
3/0	6/1	.802	391.0	370.0	374.5
4/0	6/1	.863	468.9	445.6	450.6
266.8	18/1	.909	472.5	448.7	453.8
266.8	26/7	.942	552.2	528.0	533.1
336.4	18/1	.984	568.2	541.6	547.3
336.4	26/7	1.020	668.0	641.2	646.9
477	18/1	1.114	758.0	726.6	733.3
477	26/7	1.158	898.7	867.1	873.9

A11.8 Guying, Messenger (Strand), and Shield Wire

A11.8.1 Zinc-Coated Steel (7 Strand)

Guying, messenger (strand), and shield wire is often galvanized steel (which includes extra-high strength [EHS] and high-strength [HS] Siemens Martin, and utilities grade).

Formed wire dead ends (FWDE) (also called *preforms*) are color coded and labeled. Manufacturers suggest that guy dead ends may be removed and reapplied twice after initial installation to retension guy strands. Should it become necessary to remove a guy dead end after it has been installed for a period of three months or longer, it should be replaced with a new dead end.

TABLE A11–8   Zinc Coated Steel (7 Strand)

Size	OD (in.)	Weight (lbs./1,000 ft.)	Preform Color Code	Utilities Grade Strength
5/16	.327	225	Black	6,000
3/8	.360	273	Orange	11,500
7/16	.499	388	Green	18,000
1/2	.517	510	Blue	25,000

A11.8.2 Alumaclad Steel "Type M Guy Strand"

TABLE A11–9   Alumaclad Steel "Type M Guy Strand"

Designation	AWG (equiv.)	OD (in.)	Weight (lbs./1,000 ft.)
5/16 in. MG3	3/7 AWG	5/16	142.7
10 MG	7/10 AWG	0.306	165.1
5/16 in. MG	—	5/16	171.6
11.5 MG	—	0.330	192.0
12.5 MG	7/9 AWG	0.343	206.2
3/8 in. MG	—	3/8	228.4
14 MG	—	0.363	232.2
16 MG	7/8 AWG	0.386	260.0
18 MG	—	0.417	306.6
7/16 in. MG	7/7 AWG	7/16	333.6
20 MG	—	0.444	347.5
1/2 in. MG	—	1/2	432.0
25 MG	—	0.519	474.8

A.11.9 Connector Data

A11.9.1 Ampacts

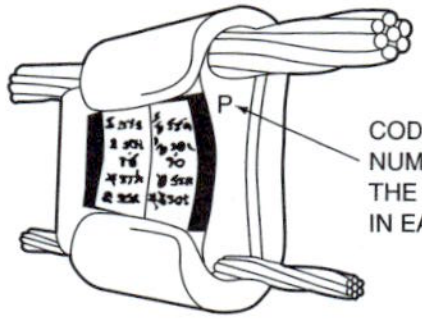

83282-1

Pub No.
Rev. 10/20/04

CODE LETTER OF
NUMBER INDICATES
THE LARGER GROOVE
IN EACH WEDGE.

MAIN LINE		YELLOW CARTRIDGE #69338-4		BLUE CARTRIDGE #65338-1				WHITE CARTRIDGE #69338-5			
		500 CU 3/4 / 477 ACSR / 556 AAC 7/8	336.4 5/8	250 CU / 266.8 ACSR / AAC 5/8	4/0 9/16	4/0 1/2	2/0 7/16	1/0 3/8 5/16	#2 9/32	#4 1/4	#6 3/16
#6	3/16	1-602031-0	6020314	602046-1	600455	600446	600446	602283-2	602283-2	602283-4	602283-4
#4	1/4	602031-9	6020313	602046-2	600456	600447	600447	602283-1	602283-2	602283-3	
#2	9/32	602031-8	6020000	602046-3	600411	600448	600403	602283	602283-1		
1/0	5/16	1-602031-9	6020001	602046-4	600458	600411	600448	600403			
2/0	7/16	1-602031-8	6020002	602046-5	600459	600458	600411				
3/0	1/2	1-602031-7	6020003	602046-6	600465	600459					
4/0	9/16	1-602031-6	6020004	602046-7	600466						
250 CU	9/16	1-602031-6	6020006	602046-7							
266.8 ACSR, AAC	5/8	1-602031-5	6020006	602046-9							
336	5/8	1-602031-4	6020007								
500 CU 3/4 / 477 ACSR / 556 AAC 7/8		1-602031-3									

All fractions are comparable steel
wire sizes.

THIS CHART TO BE USED FOR A.C.S.R., A.A.C. & STRANDED COPPER ONLY.
For solid copper, smooth body and other wire types, see master chart (GP 1931).

MULTIPLE CONNECTIONS

NO. WIRES		TYPE	GROOVE SIZE
2	1/4	strd.Alum#2 (S.B.)	2/0
3	1/4	strd.Alum#2 (S.B.)	4/0
2	1/4	strd.Alum#4 or #4 ACSR	1/0
3	1/4	strd.Alum#4 or ACSR	2/0
2	9/32	strd.Copper #2 or Alum. Std. Round	3/0
3	9/32	strd.Copper #2 or Alum. Std. Round	266.8
2	9/32	ACSR #2 Std. Round	4/0
3	9/32	ACSR #2 Std. Round	266.8
2	9/32	strd.Alum#2 (S.B.)	3/0
3	9/32	strd.Alum#2 (S.B.)	266.8
2	5/8	strd.Alum 250 (S.B.)	556.5

Tyco Electronics Energy Division
800 Purfoy Road
Fuquay-Varina, NC 27526-9349
T:800-327-6996
F:800-527-8350
www.tyco.com

Tyco Electronics Canada Ltd.
20 Esna Park Drive
Markham, Ontario L3R 1E1
Phone: (905) 475-6222
Fax: (905) 470-5271
www.tyco.com

Figure A11–1

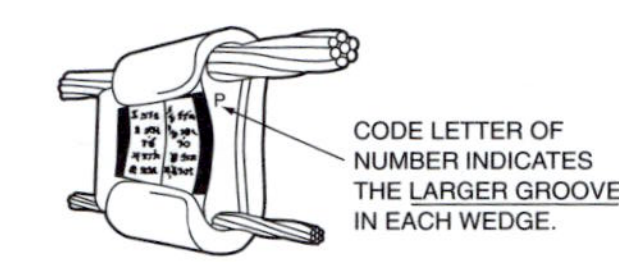

YELLOW CARTRIDGE #69338-4

MAIN LINE	795.0	795.0 CU	556.5 AAC / 636.0 AAC	477.0 / 500.0 CU	336.0
#6	1-602121-4	1-602121-4	2-602031-2	1-602031-0	602014
#4 AAC / ACSR	1-602121-3		2-602031-1	602031-9	602013
#2 AAC / ACSR	1-602121-2	1-602121-3	2-602031-0	602031-8	602000
1/0 AAC / ACSR	1-602121-1	1-602121-2	2-602031-8	1-602031-9	602001
2/0 AAC / ACSR	1-602121-0	1-602121-1	2-602031-7	1-602031-8	602002
3/0 AAC / ACSR	602121-9	1-602121-0	2-602031-6	1-602031-7	602003
4/0	602121-8	1-602121-9	2-602031-5	1-602031-6	602004
250 / 266.8 CU	602121-7	1-602121-8	2-602031-4	1-602031-5	602006
336.4 AAC / ACSR	602121-6	1-602121-7		1-602031-4	602007
500.0 CU / 477.0 ACSR	602121-4	1-602121-5	2-602031-3	1-602031-3	
556.5 AAC	602121-3	1-602121-4	2-602031-2		
636.0 AAC	602121-2	1-602121-3			
750.0 CU	602121-1	1-602121-2			
795.0	602121				

THIS CHART TO BE USED FOR A.C.S.R., A.A.C. & STRANDED COPPER ONLY.
For solid copper, smooth body and other wire types, see master chart (GP 1931).

AMPACT STIRRUPS

PART NUMBER	CABLE RANGE	BAIL SIZE
602585 (TYPE II)	#6	#2
602585 (TYPE II)	#4, #2	
600464	1/0, 2/0	#2
275436-1	1/0, 2/0	1/0
600468	2/0, 3/0	#2
600469		#2
275435-1	3/0, 4/0	1/0
602173		2/0
600463		2
602201	266.8	1/0
602502		1/0
276476-1	350.0	#2
600474		1/0
602142	336.4	2/0
602136		4/0
602047		1/0
602143	477.0	2/0
602247		4/0
602104		1/0
602248	556.5 / 636.0 AAC	2/0
602115		4/0
602147	636.0 ACSR	2/0
275074		4/0
602162	795.0	2/0
602163		4/0
602237	1033.5	4/0

Figure A11–2

A11.9.2 Insulated Serv-ens (Cooper Power)

TABLE A11–10   Insulated Serv-ens (Cooper Power)

End A				End B			
Color	Solid	Strand	ACSR	Color	Solid	Strand	ACSR
Green	#6	#8	—	Green	#6	#8	—
Blue	–4	–5 & 6	#6-6/1	Brown	#8	#10	—
Blue	–4	#5 & 6	#6-6/1	Green	#6	#8	—
Blue	#4	#5 & 6	#6-6/1	Blue	#4	#5 & 6	#6-6/1
Orange	#2	#3 & 4	#4-6/1-7/1	Green	#6	#8	—
Orange	#2	#3 & 4	#4-6/1-7/1	Blue	#4	#5 & 6	#6-6/1
Orange	#2	#3 & 4	#4-6/1-7/1	Orange	#2	#3 & 4	#4-6/1-7/1
Red	—	#1 & 2	#2-6.1-7/1	Green	#6	#8	—
Red	—	#1 & 2	#6-6/1-7/1	Blue	#4	#5 & 6	#6-6/1
Red	—	#1 & 2	#2-6/1-7/1	Orange	#2	3 & 4	#4-6/1-7/1
Red	—	#1 & 2	#2-6/1-7/1	Red	—	#1 & 2	#2-6/1-7/1
Yellow	—	1/0	1/0 –6/1	Blue	#4	#5 & 6	#6-6/1
Yellow	—	1/0	1/0 – 6/1	Orange	#2	#3 & 4	#4-6/1-7/1
Yellow	—	1/0	1/0 – 6/1	Red	—	#1 & 2	#2-6/1-7/1
Yellow	—	1/0	1/0 – 6/1	Yellow	—	1/0	1/0 – 6/1

CHAPTER 12

Preparing for Hot Work

12.1 Safe Work Principle for Hot-Line Work

Do not approach a circuit more than the minimum approach distance unless you are insulated, the facility is insulated, or personal protective grounds are installed on the facility.

12.2 Common Preparations for Hot Work

12.2.1 Choosing the Hot-Line Work Option

Risk is increased when a complex job is carried out "hot." A lot of jobs are carried out with the line in service without ever considering alternatives. The option of doing a job with the circuit isolated and grounded should at least be considered. However, some jobs are better done "hot" when balanced against the complexity of switching, grounding, and notifying customers. Customer impact and the job complexity should be decision factors when planning a job.

If	Then
Customer impact is negligible: • There are no critical customers. • There will be no unnecessary economic hardship on commercial customers (hospitals, gasoline stations, restaurants, manufacturing plants, dairy farms). • The duration of the outage will not adversely affect the customer (manufacturing plant, shopping mall, ventilation in chicken and pig farms, furnaces or air conditioning on extremely cold or hot days, sump pumps when groundwater is high).	Arrange for an outage and apply protective grounds.
The job is complex when carried out hot: • There is no approved work method. • The weather or lighting conditions are not favorable for hot-line work. • The complexity of doing the job hot does not match the needs of the customer.	Arrange for an outage and apply protective grounds.

12.2.2 Remove or Cover the Second Point of Contact

No current flow (electrical burns) will occur if contact is made with a hot circuit and no other part of a person's body is in contact with another object at a different potential. At distribution voltages, it is always possible to cover, remove, or maintain a minimum approach distance from a second point of contact. An insulated boom and bucket provide protection from earth as a second point of contact, but unless the structure, down guys, and other conductors are covered, the protection from a second point of contact is inadequate.

When working hot line from transmission-line structures, the removal of the second point of contact is not always an option. However, the greater approach distances available reduce the risk of inadvertent contact.

12.2.3 Minimum Approach Distances Apply to All Hot-Line Work

When using hot-line tools, the minimum approach distances listed in Table 12–1 apply to the length of the clear insulating section of the tool. The flashover voltage for a live-line tool is the same as it is for air. A fiberglass live-line tool may be better insulation than air, but the distance needed on a tool is an air gap between the hands on a live-line tool and the live conductor.

Rubber-glove work and barehand work should not be considered exceptions to the minimum approach distances found in Table 12–1. When in contact with a live conductor using rubber gloves or barehand techniques, the minimum approach distance still applies, but it applies to the distance between a worker and any second point of contact. When in contact with a hot conductor, a second point of contact with cover-up installed on it may be approached more closely than the distances specified in Table 12–1. Typically, a utility/employer allows "brush contact" or requires that an "air gap" be maintained with a covered-up conductor. No cover-up is available for transmission-line work.

For some very special transmission-lines circumstances, the minimum approach distance for hot-line tool work is reduced when a portable protective gap (PPG) is placed at an adjacent structure. Some live-line tool configurations, especially for transmission-line work, will have a metal fitting such as a splice or a tool-end fitting within the insulated portion of the tool. This metal will introduce electrical stresses that will require extending the minimum approach distance.

12.2.4 Block the Automatic Reclose Feature on Switchgear

When hot-line work is to be done on a circuit, the automatic reclosing feature of the source circuit breaker or recloser should be blocked from service and tagged to prevent reclosing.

- The nonreclose feature does not prevent an accident from occurring and does not guarantee that the circuit will trip out during an accident. When set in a nonreclose position, the breaker or recloser does not operate faster than normal. The nonreclose feature does guarantee that a circuit will not be reenergized once it does trip out.

- When the circuit is not automatically reenergized, the exposure time to the electrical hazard is reduced, which will allow a safer rescue effort if needed.

- Damage or injury may be limited by the reduced exposure to the high fault current that can occur during a fault.

- The risk of a restrike, which causes a switching surge (voltage surge), is eliminated if the circuit breaker is put into a nonreclose position.

12.2.5　Hazards Specific to Using Temporary Jumpers

Hazards	Barriers
Insulation failure of insulated jumpers.	Ensure that an insulated jumper has been tested as scheduled. Do a visual inspection for damaged rubber. Maintain at least a minimum approach distance from the terminations.
The current-carrying capacity of the jumper is not adequate. An arc is generated when the main conductor is cut.	Inspect the cable near the terminations for broken strands. Use an ammeter to ensure a jumper is carrying at least 30 percent of the load current before cutting a conductor or connection. The jumper conductor size should be equal to the size of the circuit conductor. The following are the typical sizes: #2 to carry 200 amps #1/0 to carry 260 amps #2/0 to carry 300 amps #3/0 to carry 350 amps #4/0 to carry 400 amps
A worker gets between a jumper and a main conductor.	Always make and break connections with a hot-line tool. Rubber jumpers with insulated clamps can be installed and removed when wearing rubber gloves. If wearing rubber gloves, you are wholly dependent on their integrity. In case of an arc being generated, you will be within the flash.
Loss of control while installing a jumper.	Once one end of a jumper is installed, the other end is hot. Use two people with two clamp sticks, or use a jumper with a parking stud so that the other end of the jumper is secured to the same terminal.

12.2.6 Using Hot-Line Rope

Reduce the risks involved with using hot-line rope by using only rope identified as hot-line rope, electrically tested for its full length and retested on a scheduled basis. A soiled, damp rope will become conductive very quickly.

- Store and transport the rope in a special container with a desiccant to absorb any moisture.

- Test the rope in the field with a portable hot-stick tester.

- Handle the rope with clean and dry work gloves.

- Clean any snatch blocks or hand-line block that the rope will run through.

- Use a tarpaulin where the rope may touch the ground.

12.3 Working on a Hot Secondary

12.3.1 Specific Hazards for Working a Hot Secondary

Specific Hazards	Barriers
Putting oneself in series with the circuit	Rubber gloves will give a person a second chance if an open point is accidently bridged.
Making contact while touching a second point of contact	Work on hot secondary voltage without rubber gloves requires avoiding a second point of contact.
An uncontrolled wire makes contact with the primary or other phases of the secondary	When the distance from the primary to the service dead end is limited, cover up the primary conductors.
Clearances between secondary bars on a pad-mount or underground transformer is close. An accidental short close to the transformer would be very explosive.	Cover up the terminals not being worked on. Work from a rubber mat. Wear rubber gloves and eye protection.

12.3.2 Explosive Flash Hazard Near a Transformer

Remember that the level of fault current at a work location will depend on the distance from the transformer and the conductor size, as shown in Figure 12–1. A dead short at the secondary terminals of a 100-kVA transformer will be greatly overloaded and generate up to 18,000 amperes. Wear eye protection when working on a hot secondary because an eye can be permanently damaged by a large flash.

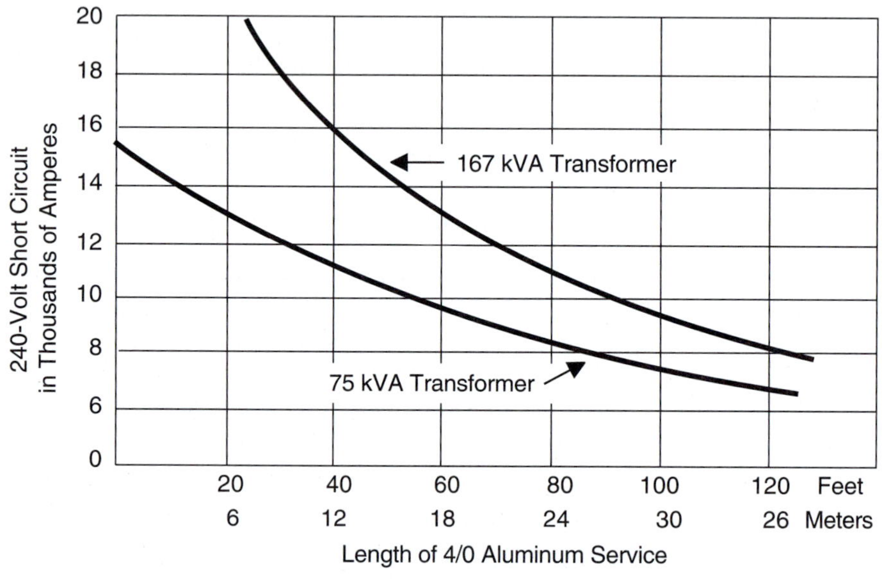

Figure 12–1   Level of fault current at various distances.

12.4 Rubber-Glove Work Preparation

12.4.1 Preparing for Rubber Glove Work

Step	Action	Details
1	Inspect rubber gloves.	• Ensure that the correct class of rubber glove (sleeve) is to be used. (see Table 12–2). • Check to ensure that the rubber gloves (sleeves) are not due for an electrical retest. Minimums: *rubber gloves,* 6 months; *rubber sleeves,* 12 months; *rubber-insulating blankets,* 12 months. Other rubber goods, such as *hose, insulator covers, and insulated jumpers* require tests upon any indication that the insulating value is suspect.

(continued)

12.4.1 Preparing for Rubber Glove Work *(continued)*

Step	Action	Details
		• The minimum distance between the top of the leather protector glove and the rolled up top of the rubber glove: Class 1 greater than 1 inch (25 mm) Class 2 greater than 2 inches (51 mm) Class 3 greater than 3 inches (76 mm) Class 4 greater than 4 inches (102 mm) • Carry out an inspection and air test on the rubber gloves. Use a mechanical inflator if possible. • Do not wear a signet-type ring with a raised surface.
2	Inspect the protective cover-up to be used.	• Rubber cover-up voltage ratings parallel the same ratings as rubber gloves. For example, the use of a class 2 rubber blanket, cover, coupler, line hose, or hood is approved for work on phase-to-phase 17,000 volts. • The voltage rating of rigid plastic cover-up is not based on the air space between the conductor and the cover. The voltage or class ratings of plastic guards are not the same as for rubber, as can be seen in Table 12–2. • A physical inspection of rubber or plastic cover-up is very effective. In almost all cases, cover-up that has failed an electrical test also had an obvious physical defect. Use only clean cover-up. • Roll rubber hose inside and outside to look for cuts, rope burns, and corona damage. Roll rubber blankets from corner to corner inside and outside while looking for damage. • Inspect plastic cover-up, such as conductor guards, insulator guards, crossarm guards, and pole guards, for cracks and cuts. The ends and lips of the covers are particularly susceptible to cracks during cold weather.

(continued)

12.4.1 Preparing for Rubber Glove Work *(continued)*

Step	Action	Details
3	Inspect insulated jumpers.	In addition to inspecting the insulation on a jumper cable, check for broken strands near the terminals. The insulation can be peeled back somewhat at each terminal for a visual inspection, or you can use the type of continuity tester used to test portable protective grounds. Ensure that the insulation of the jumper is good for the voltage being worked on. Store in appropriate bags when not in use.
4	Use ground-to-ground, extended reach, or the cradle-to-cradle rubber-glove rule, as mandated by your utility/employer.	*Caution:* A conclusion in many electrical accident investigations is "The victim was not wearing rubber gloves."
5	Plan protection against any second point of contact.	Rubber-glove work should be planned and carried out so that a worker is never totally dependent on the integrity of rubber gloves. Having this second line of defense is essential when working on a circuit with rubber gloves. • Work from an insulated aerial device, or if that is not possible, • Work from pole-mounted insulated platforms, but ensure that the top and bottom bare sections are clean, dry, and on an electrical retest schedule. • Stay at a minimum approach distance from all second points of contact or cover-up or remove the neutral, secondary wires, guys, or other phases. • If two people are working at the same structure, do not work on more than one phase at a time.

TABLE 12–1

Maximum Phase-to-Phase Voltage (Max. Phase-to-Ground Voltage)	Minimum Approach Distance Phase-to-Ground Exposure	Minimum Approach Distance Phase-to-Phase Exposure
0.05 to 1.0 kV	Avoid Contact	Avoid Contact
Up to 15 kV (8.7 kV)	2 ft., 1 in. (64 cm)	2 ft., 2 in. (66 cm)
Up to 36.0 kV (20.8 kV)	2 ft., 4 in. (72 cm)	2 ft., 7 in. (77 cm)
Up to 46.0 kV (26.6 kV)	2 ft., 7 in. (77 cm)	2 ft., 10 in. (85 cm)
Up to 121 kV	3 ft., 2 in. (95 cm)	4 ft., 3 in. (1.29 m)
Up to 145 kV	3 ft., 7 in. (1.09 m)	4 ft., 11 in. (1.50 m)
Up to 169 kV	4 ft. (1.22 m)	5 ft., 8 in. (1.71 m)
Up to 362 kV	8 ft., 6 in. (2.59 m)	12 ft., 6 in. (3.8 m)
Up to 550 kV	11 ft., 3 in. (3.42 m)	18 ft., 1 in. (5.50 m)
Up to 800 kV	14 ft., 11 in. (4.53 m)	26 ft. (7.91 m)

TABLE 12–2

Class	Phase-to-Phase kV	Phase-to-Ground kV
2	14.6	8.4
3	26.4	15.3
4	36.6	21.1
5	48.3	27.0
6	72.5	41.8

12.4.2 Rubber-Gloving Hazards and Barriers

Rubber-Glove Hazards	Barriers
Overreaching exposes an uncovered part of the body to an electrical contact.	1. Work from a position where a slip will not cause an uncovered part of the body to make contact—for example, from a position below the conductor. 2. Wear rubber sleeves to reduce the risk of contact due to overreaching.

(continued)

Rubber-Glove Hazards	Barriers
Untying and tying-in conductor, because of the following: 1. The length of the tie wire. 2. Possible puncture of the rubber glove by the sharp end of the tie wire.	Bunch up the tie wire in the palm of your rubber glove when untying.
Installing and removing preformed dead ends or armor wrap. The preforms for larger conductors are long enough to contact other phases or objects.	For large conductor, use a dead-end clamp (shoe). If it is necessary to remove a long preform, make frequent cuts.
Temporarily removing rubber gloves for fine work or to cool off increases the probability of an inadvertent contact.	A mental lapse can occur anytime, including when rubber gloves have been removed. Tell someone working with you of your intention.
Working on more than one phase at a time. The other phase is a potential second point of contact.	Two people working on the same structure must not work on different phases at the same time. One can become a second point of contact for the other.
Working on adjacent structures at the same time.	When conductors are moved on adjacent structures at the same time, there must be communication between the two poles.
Small conductor can break while working on it.	Conductors such as #6 copper, #4 ACSR, and #8A copperweld, or smaller, should not be worked hot.
Long, metallic tools can bridge or short the rubber-glove cuff.	Long-handled tools, such as a bolt cutter, press, hoist, or hammer, should have nonmetallic handles.
Improvising the temporary placement of a conductor while replacing a crossarm or other support can lead to poor clearance between conductors and to workers.	Use tools such as an auxiliary mast to support conductors while replacing a crossarm or other support items.

12.5 Hot-Line Tool Work Preparation

12.5.1 Preparing to Work with Hot-Line Tools

Step	Action	Details
1	Identify the length of tools required.	1. Hot-line tools must be long enough to provide a minimum insulated working distance between any part of a worker and the energized conductor or part being worked on (see Table 12–1). Ensure that the clear insulation of conductor support tools, such as link sticks, strain carriers, and insulator cradles, are at least as long as the insulator string or the minimum distance specified in Table 12–1 for the operating voltage. 2. The weights and tensions of conductors to be lifted or pulled can be calculated as described in Chapter 15. Lifting conductor at a hilltop or at a corner can increase the weight and tension on the sticks very quickly. 3. Check the working-load limit of the tool configuration to be used. Check the working-load limit of the components to be used, such as the saddles, tongs, link sticks, and snubbing bands. 4. Check that the work method will allow compliance with the minimum approach distance. 5. Check the conductor attachments at adjacent structures.
2	Inspect hot-line tools.	1. Check to see that the tools have had the minimum OSHA required examination, that is, cleaning, repair, and electrical test within the last two years. 2. Use hot-line tools that are kept clean and dry and in a protective container when not in use. 3. A physical inspection of hot-line tools is very effective. In many cases, a stick that fails an electrical test also has an obvious visual physical defect. Physical damage to the tool, such as surface scratches or a deformed or popped rivet, can allow the moisture to enter.

(continued)

12.5.1 Preparing to Work with Hot-Line Tools *(continued)*

Step	Action	Details
		4. Hollow tools are more susceptible to contamination and may need more frequent electrical testing.
3	Wipe or clean hot-line tools.	**1.** Clean and dry your tools with an agent specifically made for this purpose—that is, an agent that evaporates quickly and leaves a treatment on the surface that repels and beads water. **2.** Damp and dirty work gloves are the most probable source of surface contamination, especially if both ends of a stick have a tool on it and the worker is using both ends.
4	Check saddles, lever lifts, and chain tighteners.	**1.** Use saddles that are stored with the holding clamp in the closed position when not in use. **2.** Check for distortions, loose bolts, and loose rivets. **3.** When the wire-tong saddles are installed on the hot-line tool, an additional 180-degree turn on the wing nut may be applied after hand tightening. **4.** Use rope blocks to make lifts with the lifting tong. As a second barrier if rope blocks are removed, tie a rope sling between the saddle and the butt ring. **5.** After a lever lift is flipped into position, tie it up to prevent it from dropping during the movement of the conductor.
5	Check ropes and blocks.	**1.** Use ropes and rope blocks that are kept exclusively for hot-line tool work.
6	Protect against any second points of contact.	**1.** Work from an insulated pole-mounted platform or an insulated aerial bucket. **2.** Cover up or remove the neutral, secondary wires, guys, and other phases. **3.** Do not work on more than one phase at a time.

12.5.2 Hot Stick Rigging Configurations

12.5.2.1 Typical Working-Load Limits on Hot-Line Tools

Check specifications from manufacturers and from your utility/employer for actual working-load limits when using hot-line tools (see Table 12–3).

TABLE 12–3   Working-Load Limits for Hot-Line Tools

Tool	Type or Size	Working-Load Limit
A wire tong (pole) is designed for compression loading. All side pulls must be balanced with an opposite side pull. Typically, for a vertical lift the minimum distance between saddles is 5 ft. (1.5 m) and the maximum unsupported length above the top saddle is 6 ft. (1.8 m).	2.5 in. (6.4 cm) 3 in. (7.6 cm)	The limiting factor for the working-load limit for a lifting pole is the saddles holding it and any horizontal strains being applied to it.
Pole saddles must hold the weight of the lifting tong, the conductor weight, and the pull exerted by the fall line on a set of blocks. For a heavy conductor, hang the set of blocks in a separate snubbing band or sling instead of the saddle.	Saddle without extension Saddle with extension	800 lbs. (360 kg) 600 lbs. (270 kg)
Lever lifts are used for the heaviest loads to be lifted.	Single-lever lift Double-lever lift	1,000 lbs. (450 kg) 1,500 lbs. (680 kg)
The allowable tension on a link stick depends on the diameter.	1.25 in. (3 cm) 1.5 in. (4 cm)	3,500 lbs. (1,580 kg) 6,500 lbs. (2,950 kg)

(continued)

TABLE 12–3 Working-Load Limits for Hot-Line Tools *(continued)*

Tool	Type or Size	Working-Load Limit
The working-load limits for suspension sticks and strain carrier poles used to lift	Typical heavy-duty suspension link stick	6,500 lbs. (2,950 kg)
or tension conductors on transmission lines vary and should be positively identified and checked out before using.	Typical strain carrier pole	7,500 lbs. (3,400 kg)
Roller link stick	1¼ in. (3 cm)	1,000 lbs. (450 kg)
The allowable tension on a snubbing band is shown per ring and the total for the complete unit.	—	500 lbs. (230 kg) per ring, the total load not to exceed 1,000 lbs. (454 kg)

12.5.2.2 Calculating the Load on Hot-Line Tools

Most hot-line tool work on transmission lines uses the tools in a straight tension mode, and the maximum safe working load can be determined from the ratings of the individual tools. For a configuration such as the one shown in Figure 12–2, some measuring and calculations are needed to determine the maximum safe working load.

The compression load on lifting tong B =

$$\frac{Length\ of\ B \times Conductor\ Weight}{Length\ of\ B} \quad or \quad \frac{B \times W}{C}$$

The tension on the holding tong A =

$$\frac{Length\ of\ A \times Conductor\ Weight}{Length\ of\ C} \quad or \quad \frac{A \times W}{C}$$

where A = length of holding tong between the conductor and the saddle
  B = length of holding tong between the conductor and the saddle
  C = distance between the saddles
  W = weight of the conductor in pounds or kilograms

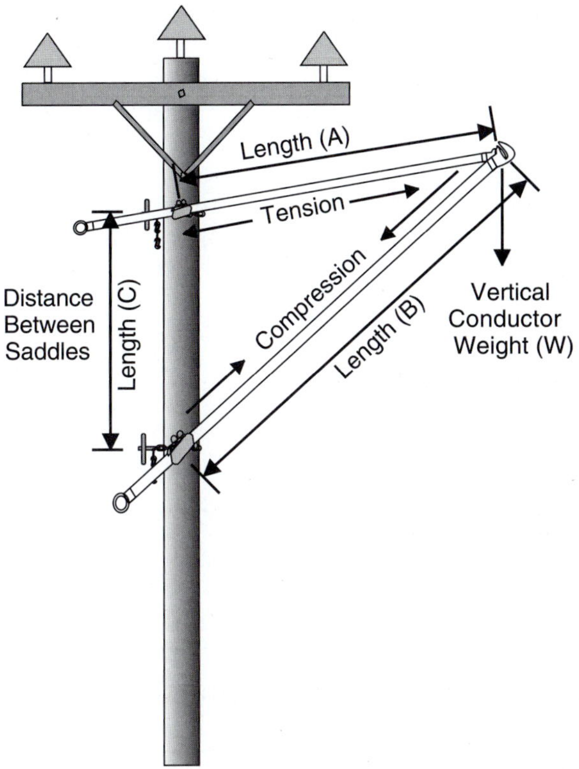

Figure 12–2   Calculating load on live-line tools.

Remember that the maximum height a conductor can be lifted occurs when the lowest point of sag reaches an adjacent structure. At this point the conductor weight has doubled. When planning to lift a conductor with an aerial device or with hot-line tools, multiply the weight to be lifted by a factor of 2. When the conductor to be lifted is at a corner structure, the counteracting force of a link stick down to a temporary rope guy holding the bisect tension of the conductor must also be added to the weight to be lifted.

12.5.2.3 An Example of Hot-Line Tool Calculations

Figure 12–3 shows two different tool configurations. If the conductor weight is given at 200 pounds, Table 12–3 shows the compression load on the lifting tong and the tension on the holding tong for the initial position and the final position for Pole A and Pole B. Note that the distance C can be critical to the tool loading.

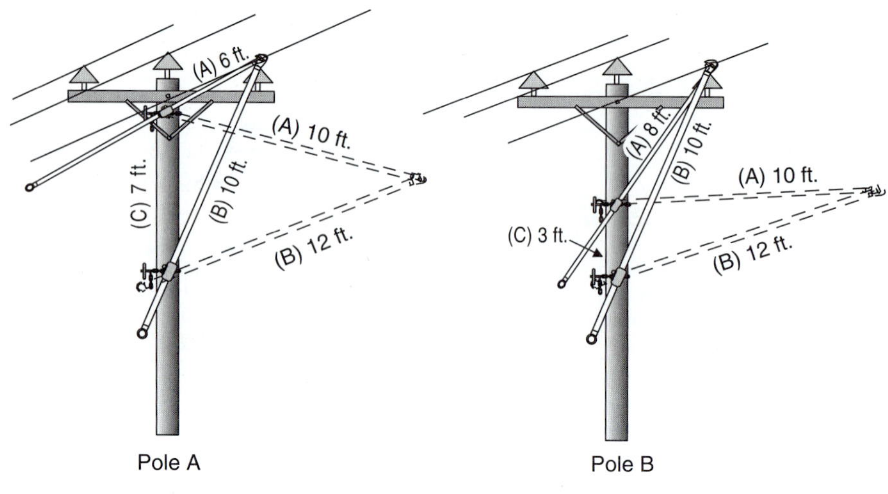

Figure 12–3   Example calculations.

TABLE 12–4   Example Calculations

	Pole A		Pole B	
	Initial Position	**Final Position**	**Initial Position**	**Final Position**
Load on Lifting Tong = $\frac{B \times W}{C}$	$\frac{10 \times 200}{7} = 286$ lbs.	$\frac{12 \times 200}{7} = 343$ lbs.	$\frac{10 \times 200}{3} = 667$ lbs.	$\frac{12 \times 200}{3} = 800$ lbs.
Load on Holding Tong = $\frac{A \times W}{C}$	$\frac{6 \times 200}{7} = 170$ lbs.	$\frac{10 \times 200}{7} = 286$ lbs.	$\frac{8 \times 200}{3} = 533$ lbs.	$\frac{10 \times 200}{3} = 667$ lbs.

12.5.2.4 Hazards and Barriers of Working with Hot-Line Tools

Specific Hazards	Barriers
Encroaching on the minimum approach distance on the tool.	Hand stops mounted just beyond the minimum limit of approach can serve as a barrier.
	Some utilities require rubber gloves in addition to the hot-line tool.
Using a contaminated damp work glove on a tool can leave a conductive film.	The most likely cause of a tool failure is surface contamination. Use clean, dry work gloves.

(continued)

12.5.2.4 Hazards and Barriers of Working with Hot-Line Tools
(continued)

Specific Hazards	Barriers
Tools sliding through saddles after weight is applied.	Use rope blocks to make the lift. Leave the rope blocks in place to hold the weight or tie a sling between the saddle and the butt ring as a backup to saddles.
Shorting out a tie wire or preform to the structure.	When untying insulators, cut the tie wire short enough so that it cannot reach any part of the structure. Consider alternative hardware for dead-ending or clamping in.
Hot-line tools left on the circuit are susceptible to tracking and leaving a carbon trail.	Hot-line tools, especially components such as extension arms, can be left on a circuit for up to 3 weeks in a nonpolluted area. Look for visible signs of tracking before handling them.
Misuse of tools can cause a loss of control.	Never use a lifting tong in a tension mode as a link stick. Double check the attachment of the saddles and snubbing bands before securing a hot conductor to it. There is no backup when a snubbing band lets go.
At transmission-line voltages, there is some suspicion that high winds can cause a decrease in air pressure on one side of the tool and the low pressure on that side of the stick causes a flashover.	As a prudent measure, avoid hot-line tool work on transmission lines during high-wind conditions or increase the maximum approach distance on the tools.

12.6 Barehand Work

12.6.1 Using an Aerial Device

An aerial device used for barehand work will have a boom contamination meter circuit built where a meter can provide continuous monitoring for leakage. The empty buckets are raised so that some part of the metal grid makes contact with the line. The boom is left in contact for a minimum of 3 minutes. The leakage current is monitored and may not exceed 1 microampere per phase-to-ground kilovolt. For example, a 230-kilovolt line is about 133 kilovolts to ground. The maximum leakage current allowed is 132 microamperes, but, depending on the type of meter, the alarm on the monitoring circuit may sound at 70 microamperes. The meter readings should be recorded in a log because one of the best measures is to see if a trend is developing.

12.6.2 Minimum Approach Distance for Barehand Work

When bonded on, minimum approach distances similar to those specified in Table 12–5 are kept from everything not already bonded to the conductor.

TABLE 12–5   Barehand Minimum Distances

Voltage Range Phase-to-Phase Kilovolt	Phase to Ground		Phase to Phase	
	Feet–Inches	*Meters*	*Feet–Inches*	*Meters*
30.0 to 35.0	2–4	0.7	2–4	0.7
35.1 to 46.0	2–6	0.8	2–6	0.8
46.1 to 72.5	3–0	0.9	3–0	0.9
72.6 to 121.0	3–4	1.0	4–6	1.4
138.0 to 145.0	3–6	1.1	5–0	1.5
161.0 to 169.0	3–8	1.2	5–6	1.7
230.0 to 242.0	5–0	1.5	8–4	2.5
345.0 to 362.0	7–0	2.1	13–4	4.1
500.0 to 552.0	11–0	3.4	20–0	6.1

12.6.3 Conductive Clothing and Footwear

Conductive clothing serves as a shield around the worker, keeping the electric field effect outside of the shielding. Coveralls, work gloves, socks (not always worn), and a parka-style hood are made with conducting material. The hood is worn over the hard hat and extends forward somewhat to help shield the face from annoying sparking along the skin. Coveralls have a tail that is used to bond to the metal grid in the bucket. Conducting sole boots are worn, and conducting straps bonded to the boots are bonded to the suit with leg clips.

After the suit is put on, an ohmmeter is used to check the continuity of the suit between the tail and the leg cuffs, the gloves, the hood, and the boot soles. All the measurements should read less than 100,000 ohms.

12.6.4 Some Specific Barehand Work Hazards and Barriers

Specific Hazards	Barriers
Electrical failure of equipment	• Watch for hot-line rope that has become somewhat contaminated with use, especially when humidity is high. • Ensure that there is a check valve and a clean atmospheric vent valve in the hydraulic lines that extend over 35 ft. (10.7 m) high to prevent a conductive partial vacuum from forming. The hydraulic motor should not be shut down during barehand work. • When metal is in series with rope or link sticks (a floating electrode), a longer minimum approach is required.
Flashover of insulated tools	• Wet, dirty work gloves can contaminate the surface of a hot-line tool, so use clean, dry work gloves. • Avoid barehand work during high winds. There is some suspicion that the slight reduction in air pressure on the lee side of a boom or tool can reduce the insulating value. • Reduce exposure to a voltage surge. Do not work barehand during electrical storms. • Reduce exposure to a voltage surge due to switching, and have the reclose function of the source breaker blocked.
Flashover of an aerial device boom while doing the contamination tests	There are conditions, such as high humidity, when the meter readings are not acceptable. If the readings are going down while the boom is in contact, it is probably leakage current that is drying out the boom. Do not use leakage current to dry the boom. This leakage current can leave conductive carbon tracking and lead to a boom flashover.
Encroaching on the minimum approach distance	Use a dedicated observer to watch for encroachment on the minimum approach distance. The most likely encroachment has been with the heel (elbow) of an aerial device. Measure a hot-line ladder or platform to ensure it will provide a sufficient length of clear insulation for the voltage to be worked on.

(continued)

12.6.4 Some Specific Barehand Work Hazards and Barriers *(continued)*

Specific Hazards	Barriers
Placing bonding clamps on each side of a hot joint or connector (The typically small bonding wire may carry current and burn off.)	Install a jumper across a hot joint or connector, using hot-line tools, before moving in to bond on.
Forgetting to install a jumper or taking an ammeter reading on a jumper before cutting	Forgetting to install a jumper can happen during a repetitious job. Considering the consequences, a crew should use a written form similar to the "Switching Order." Use an ammeter to ensure that the jumper is carrying at least 30 percent of the load before cutting or opening a conductor barehand.
Shorting out a portion of the insulator string with the buckets or body	When using slings and hoists near suspension insulators, it is possible to short out some insulators in a string. If some of the insulators are already defective, the risk is compounded. Test the insulators. If more than 20 percent of the insulators have no insulation value, surge-limiting devices can be installed prior to barehand replacements, or insulators can be replaced by other than barehand procedures. For work at the energized end of an acceptable insulator string, a bucket or worker should be positioned so that no more than 10 percent of the insulator string is shorted out by the buckets, worker's body, or tools.
Conductive objects between a hot circuit and another phase or structure that form afloating electrode (which can reduce the insulating value across an air gap)	A person in a conductive suit, a metal grid in the buckets, or two link sticks tied together form floating electrodes. A person in a conductive suit working between a tower and conductor on an insulated ladder will be a floating electrode. These objects can give off an electrical discharge that reduces the air-gap distance needed to prevent a flashover. Increase the maximum approach distance when a floating electrode is introduced.

(continued)

12.6.4 Some Specific Barehand Work Hazards and Barriers *(continued)*

Specific Hazards	Barriers
Contacting an object before bonding to it	It is probably not necessary to point this out to an experienced barehand powerline worker, but there is a need to bond or bridge to any metallic objects to be touched or brought into the work zone.

12.7 Hazards Specific to Transmission Line Hot Work

12.7.1 Voltage Surge on Transmission Lines

Options to Reduce the Magnitude of a Voltage Surge Due to Lightning or Switching

To encourage a surge to flash over on a structure where work is not being carried out, consider the following options:

- Surge gap distance at the line terminals can be reduced temporarily for the duration of the work.
- Surge arrestors attached at the line terminals will reduce the magnitude of a voltage surge.
- A temporary portable protective gap (PPG) can be installed at an adjacent structure.
- Some insulators can be shorted out at an adjacent tower.

Options to Reduce the Probability of a Voltage Surge

To reduce the likelihood of a voltage surge, consider the following options:

- Block the automatic reclose feature on all circuit breakers connected to the circuit. This will prevent switching surges after the line trips out.
- Ensure that no relay testing is being done on the circuit while hot-line work is in progress.
- Do not work when lightning is in the vicinity. The controlling station staff for transmission lines should have access to an information system that notifies them of impending electrical storms.

12.7.2 Testing Insulators for Breakdown

When hot-line work (especially barehand) is planned with or near existing suspension-type insulators, each insulator in the string should be tested. Transmission-line insulators can be tested while the line is in service with an insulator tester. Figure 12–4 shows how one type of insulator tester is slid up and down on a hot stick suspended parallel to an insulator string. This tester takes advantage of the fact that the electrical field surrounding a defective or shorted insulator decreases. It works on polymeric and porcelain/glass insulator strings.

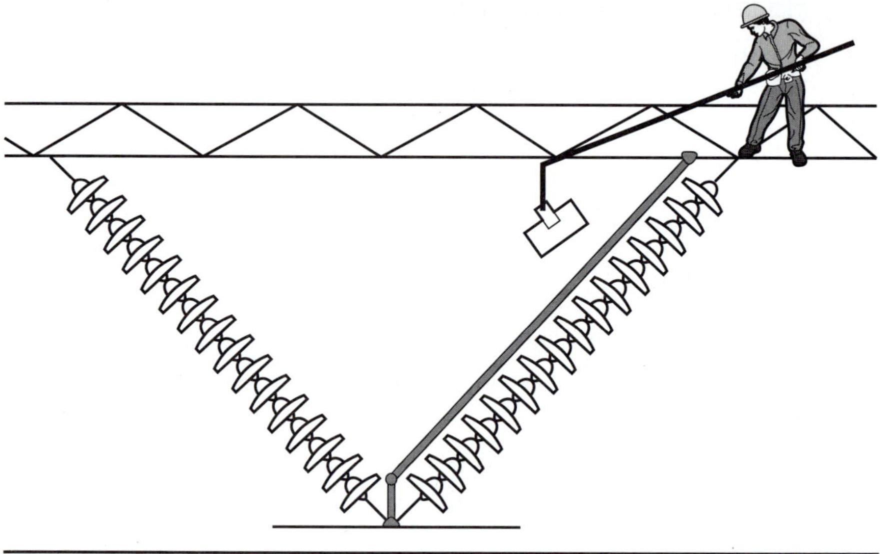

Figure 12–4   Testing insulators.

CHAPTER 13

Installing and Removing Revenue Meters

13.1 Specific Hazards Installing and Removing Meters

When a socket-connected meter (S-base) is plugged into a meter base with the top source lugs alive, installing or removing the meter is like closing or opening a switch. A short circuit or heavy load can cause an explosive flash.

The live installation or removal of a three-phase self-contained meter on a service rated at more than 240 volts should be avoided because the higher voltage combined with a potentially high fault current is extremely explosive.

Specific Hazards	Barriers
A high explosive fault can occur at a meter installation, especially at an installation fed from a large transformer and a short run of large service conductors. Faults include broken meter lugs and faulty customer wiring.	Conduct tests as described in section 13.1.1, section 13.1.2, or section 13.1.3. Use a meter installer and removal tool (Figure 13–1) on S-base meters to reduce the risk of injury from explosive faults. Wear eye protection and flame resistant clothing.

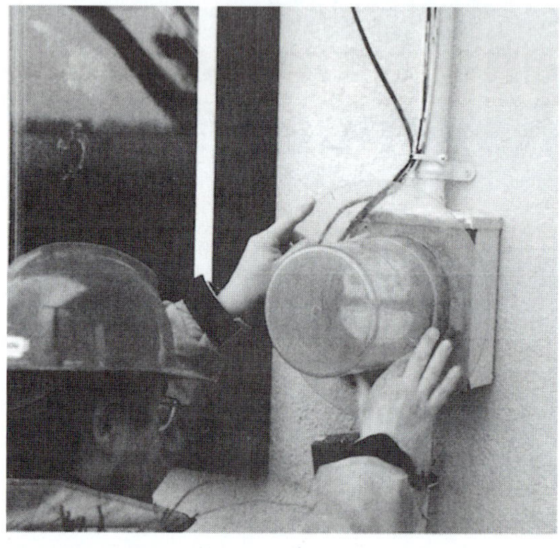

Figure 13–1   A meter installer and removal tool.

Any fault or excessive load when installing or removing a 480/277 or 600/348 volt self-contained meter alive can result in an extremely explosive flash.	Avoid installing or removing a self-contained meter on services of more than 240 volts. This type of service often has a main switch on the source side of the meter. The first option should be to isolate the service at the switch. If it is necessary to install or remove a 480/277 or 600/348-volt self-contained meter alive, use a meter installer and removal tool (Figure 13–1) and wear a full face shield, rubber gloves, and flame-retardant clothing.
Dropping an excessive load can cause an explosive fault while removing a socket-based meter.	Check the load on the meter, as per section 13.1.4. If the load is greater than 10 kilowatts, do not remove the meter under load. Open the transformer, get access to the customer's breaker, or reduce the customer's load.

(continued)

(continued)

Specific Hazards	Barriers
	Use a meter installer and removal tool (Figure 13–1) on S-base meters to reduce the risk of injury from explosive faults. Wear eye protection and flame-retardant clothing.
A high voltage is in the meter base when a transformer meter is removed.	Unless an automatic shorting device is in the meter base, short out the current transformer (CT) before removing the transformer-rated meter.
Striking or pounding on a glass cover can break the glass and cause severe cuts.	Use a meter installer and removal tool (Figure 13–1) on S-base meters to reduce the risk of injury from breaking glass.
A third party makes contact with an unprotected meter base.	If the meter base (socket) is to be left energized, install a meter or cover and seal to protect the public before leaving the site.

13.1.1 Insulation Tester Method

A 500-volt insulation tester, which is like an ohmmeter, can be used at the meter base to test for a short circuit.

If the customer's main disconnect is open, the readings should be as follows:

- L1 to L2 = infinity

- L1 to neutral = infinity

- L2 to neutral = infinity

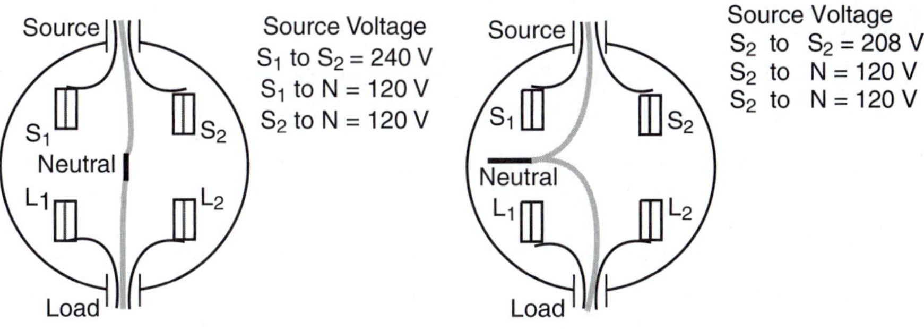

Figure 13–2   (a) A 120/240-volt single-phase three-wire meter base. (b) A self-contained 120/208-volt single-phase three-wire meter base.

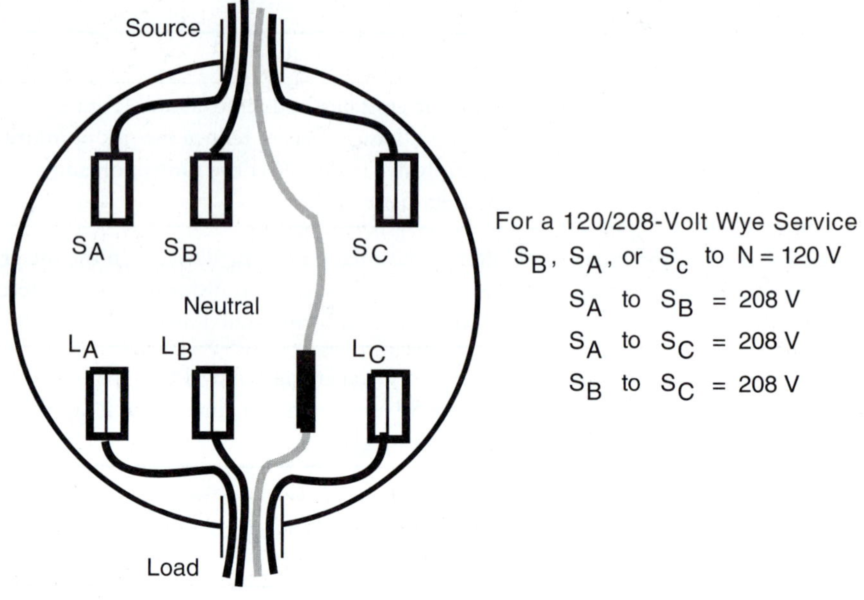

For a 120/208-Volt Wye Service
S_B, S_A, or S_C to N = 120 V
S_A to S_B = 208 V
S_A to S_C = 208 V
S_B to S_C = 208 V

Figure 13–3 A 4-wire meter base.

If the customer's main disconnect is open, the readings should be as follows:

- L_A to L_B = infinity
- L_A to L_C = infinity
- L_B to L_C = infinity
- Each of L_A, L_B, and L_C to neutral = infinity

13.1.2 Voltmeter Method

Single-Phase Meter Base

With the source side alive and the customer breaker open, any voltage reading indicates a problem that must be investigated (see Figure 13–2).

- No voltage, L1 to L2 to neutral will confirm that the service is not fed back from another source.

- No voltage, S1 or S2 to L1 and L2 will confirm that there is no load or a load wire shorted.

- No voltage, S1 to L2 with a temporary (fused) jumper between L1 and the neutral indicates that there is no short between the two load wires.

Three-Phase Meter Base

With the source side alive, customer breaker open, any voltage reading indicates a problem that must be investigated (see Figure 13–3).

- No voltage, L_A to L_B to L_C to neutral will confirm that the service is not fed back from another source.

- No voltage, from S_A, S_B, or S_C to each of L_A, L_B, or L_C will confirm that there is no connected load, the service is alive from another source, or a load wire is shorted to the neutral or ground.

- No voltage, S_A to L_B and L_C, with a temporary (fused) jumper between L_A and the neutral indicates that there is no load or short between the three load wires.

13.1.3 The Super Beast Method

Checking a 120/240-Volt Service with Super Beast

The Super Beast is an instrument that can be used to check out a service at the meter base. It imposes a 20-ampere artificial unbalanced load on the service at the meter base. This will test the customer wiring from the meter base to an open main breaker and will test the source-side service wires. A voltage drop of 5 to 7 volts is an indication of trouble.

A newer digital Mega Beast imposes an 80-ampere unbalanced load. This unit will find the more elusive faults more quickly.

How to Use the Super Beast

Job steps

1. Remove the kilowatt-hour meter and inspect the meter base. Connect the green clip to the neutral and insert the Super Beast into the meter base.

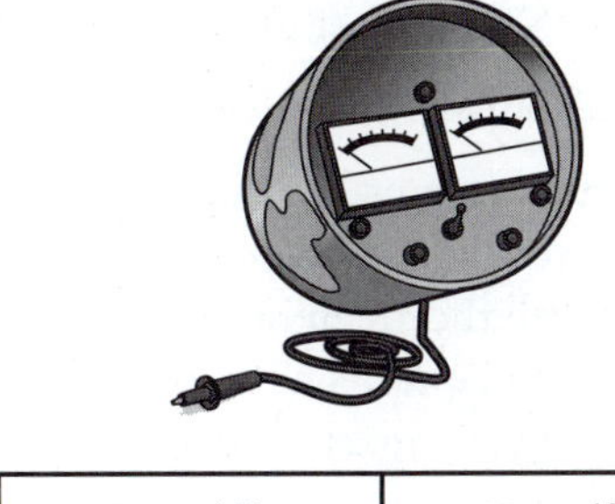

2. Push the switch to the left and read the meters.

   If the left meter reads 120 volts, *then* the left conductor is okay.

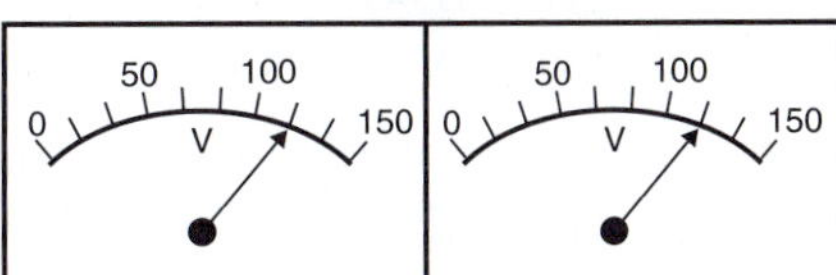

Figure 13–4   Meter readings on Super Beast.   *(continues)*

If only the left meter drops, *then* the left conductor is open.

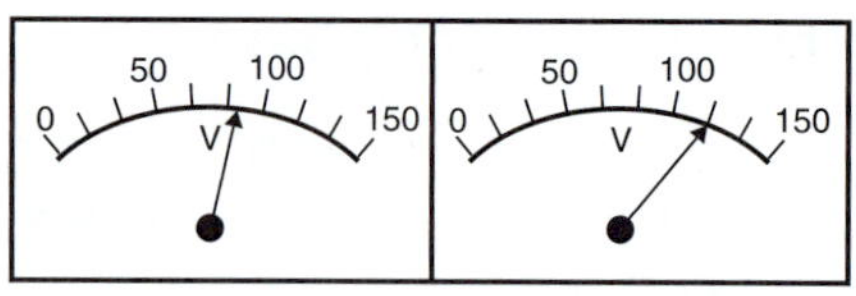

If the left meter drops and the right meter increases, *then* the neutral is open.

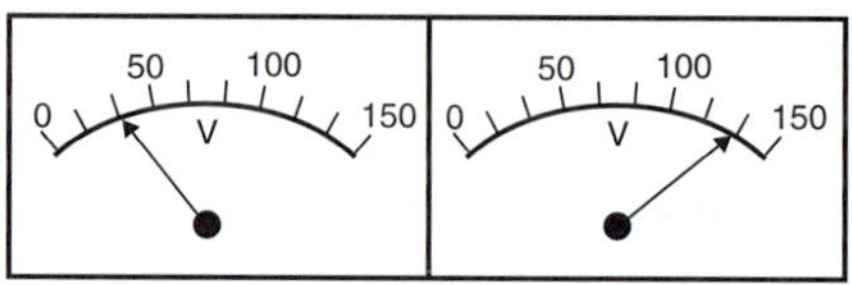

3. Push the switch to the right and read the meters.

If the right meter reads 120 volts, *then* the right conductor is okay.

If only the right meter drops, *then* the right conductor is open.

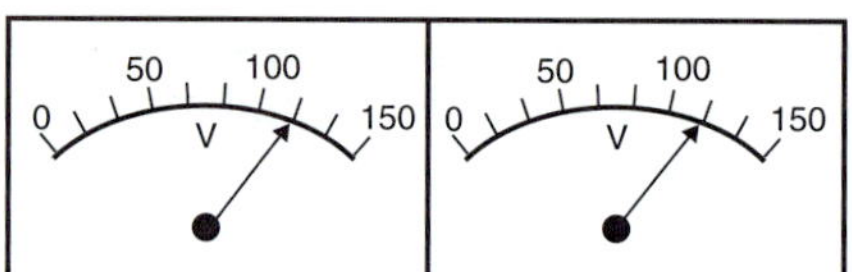

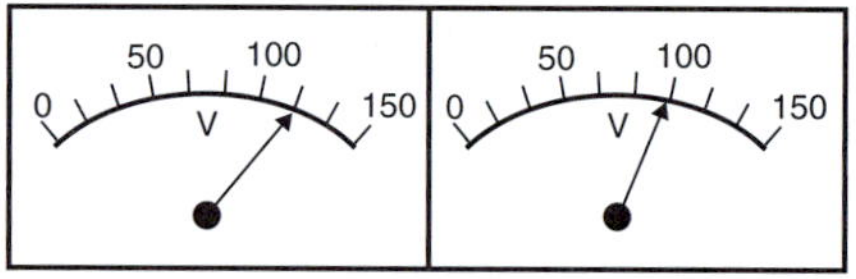

If the right meter drops and the left meter increases, *then* the neutral is open.

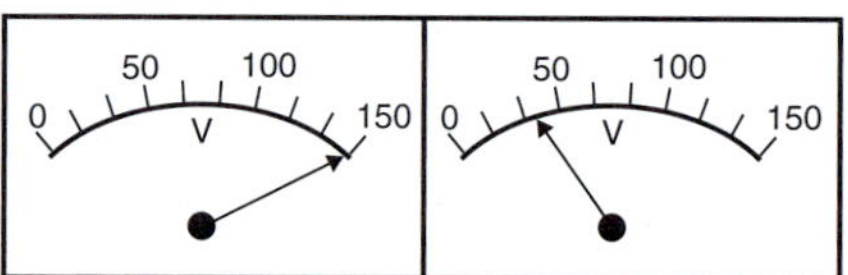

Figure 13–4 Continued

Checking a Three-Phase Wye 120/208 Service with Super Beast

- The Super Beast cannot be used on voltages higher than 240 volts.

- The Super Beast is not designed for use on a three-phase delta circuit.

A portable meter-base adapter is used between the Super Beast and the three-phase meter base. Connect the meter-base adapter to the back of the Super Beast. Install the green neutral clip of the Super Beast to the neutral bolt of the meter-base adapter. The Super Beast is not designed for use on a three-phase delta circuit. Connect the clips of the meter-base adapter to the neutral and S_A and S_B (**Note:** Super Beast instructions use the term "line" instead of source.)

1. Take readings on S_A and S_B. The meter-reading interpretation is the same as for the preceding 120/240-volt single-phase procedure.

2. Move one clip lead from either S_A or S_B to S_C and again interpret the meter readings as for the 120/240-volt single-phase procedure.

13.1.4 Check Load on a Meter

An uncontrollable electrical flash can occur when a socket-based meter is removed under load. Some utilities allow a single-phase meter to be removed if the load is 10 kilowatts or less. Table 13–1 can be used to determine whether a load is more or less than 10 kilowatts. Count the number of disk revolutions in 30 seconds; with reference to the meter (Kh), a number of disk revolutions higher than shown on the chart means that the load is more than 10 kilowatts.

TABLE 13–1   Checking Load on a Meter to Allow Safe Removal

Kh	Time in Seconds	Disk Revolutions
0.36	30	231
0.6	30	138
0.66	30	126
0.72	30	115
2.0	30	41
3.0	30	27
3.33	30	25
3.6	30	23
6.0	30	14
7.2	30	11

For example, the disk of a meter with a Kh of 7.2 will rotate 11 times in 30 seconds to register a load of 10 kilowatts. When the load is more than 10 kilowatts, the customer's main disconnect should be opened before removing the meter.

13.2 Making Metering Connections

13.2.1 Metering Connection Principles

The connections to a meter follow certain principles. Variations depend on the manufacturer and regulatory agencies. Drawings should be consulted when making connections to a three-phase meter because a mistake can lead to awkward circumstances for a utility.

- For each phase, the source wires to the current coil and the voltage coil must be the same polarity.

- The load wires from the potential coil must have a polarity opposite the polarity of the current coil. In other words, depending on the type of service and meter, the load side of the potential coil must be connected to another phase or to the neutral.

- Each current coil must be connected in series with the load.

- Each potential coil must be connected in parallel, like a voltmeter.

- The color coding of the wiring must be followed.

- The three phases must be in a proper phase rotation.

13.2.2   Electronic Metering

The electronic (static or solid state) revenue meter will eventually take over from the electromagnetic meter. A multitude of measurements can be programmed into an electronic meter. It is also more suitable for automatic meter-reading systems. In an electronic meter, the current and voltage act on solid-state (electronic) metering elements to produce an output pulse. A pulse initiator produces contact closures (pulses) proportional to the watt-hours being measured.

The electronic meter illustrated in Figure 13–5 is a three-phase, four-wire, transformer-rated electronic meter. A typical electronic meter can be programmed to carry out many functions:

- It can be a multirate meter with more than one readout, each becoming operative at a specified time corresponding to a different tariff rate.

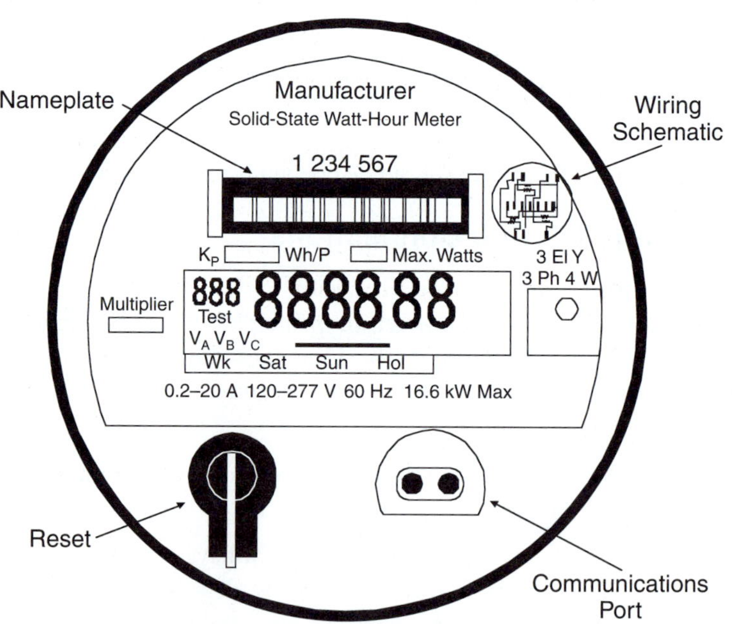

Figure 13–5   A three-phase, 4-wire, transformer-rated electronic meter.

- It can be used to check the voltage and current in each phase, the power factor on each phase, and the instantaneous kilowatt and kilovolt-ampere for each phase.

- The meter can check itself for proper function, much like a computer checks itself out. It also can check itself for proper installation, such as phase rotation.

- It can register the time of use, taking into account holidays and weekends.

- It can register kilowatt demand and kilovolt-ampere demand.

- It is easily adapted to all the benefits associated with automatic meter reading by access through the communications port.

13.3 Current Transformers

13.3.1 High-Voltage Hazard with Current Transformers

A current transformer is like other transformers:

$$\textit{The input } V \times I = \textit{The output } V \times I$$

When the current is stepped down, the voltage is stepped up at the same ratio. The voltage would be expected to be high at the terminals, but when there is a load, such as a meter or relay, a voltage drop brings the voltage to a safe level. When the load is removed, the voltage at the secondary terminals and at terminals where relays are installed will be very high. This high voltage can puncture the insulation of the current transformer, and it is a dangerous shock hazard to any person in contact with the secondary circuit.

Therefore, before a load is removed from an energized current transformer, the secondary terminals must be shorted out. Typically, a shorting device at the secondary terminals is part of the current transformer design and can be closed safely while the load is still on the secondary. Many different types of current transformer designs are in use. If you are uncertain about how to short out the secondary, get help. The meter base for a socket-mounted, transformer-rated meter often has an automatic shorting device that will close the metering circuit when the meter is removed.

13.3.2 Three-Wire Current Transformer

One current transformer (CT) can be used with a three-wire, 120/240-volt service. Both 120-volt conductors go through one CT. To have the *current* in the two opposite-polarity 120-volt legs go through the CT in the same direction, the two wires must go through the CT from opposite directions. The current from the two legs are added together to double the secondary current. For example, a 200/5 CT becomes a 200/10 CT with this setup. A three-wire CT used on a 120/240-volt service has two primary windings and one secondary winding. In Figure 13–6, the three-wire CT is actually a two-wire CT with two wires going through it acting as two primary windings. The secondary current from the CT represents the sum of the two 120-volt legs of the service.

The type of service shown in Figure 13–7 is a 120/240-volt service metered at a customer transformer pole using a three-wire current transformer.

12.3.3 A Multiplier on a Current Transformer

When an instrument transformer is used, the meter is reading and registering a representative voltage and/or current, and, therefore, the reading must be multiplied by a meter multiplier to obtain the actual status of the energy used.

The ratio of the current transformer and voltage transformer (VT) must also be calculated into the total. The VT multiplier is always equal to the transformer ratio. The CT multiplier depends on how the CT is wired. A bar-type CT multiplier is always equal to the transformer ratio. When one wire goes straight through a donut CT, the multiplier is the same as the transformer ratio. For example, a CT with a ratio of 400:5 (80:1) will have a multiplier of 80.

A CT multiplier can be reduced by looping the primary through the CT more than once. For example, when the primary is looped through the CT twice (as shown in Figure 13–8), the secondary current is doubled. A 400:5 CT will become a 400:10 CT (200:5 or 40:1), which changes the multiplier to 40. The secondary of a 400:5 CT is rated to 5 amperes. Therefore, a double-looped primary should be used only where the primary is 200 amperes or less to limit the secondary to 5 amperes.

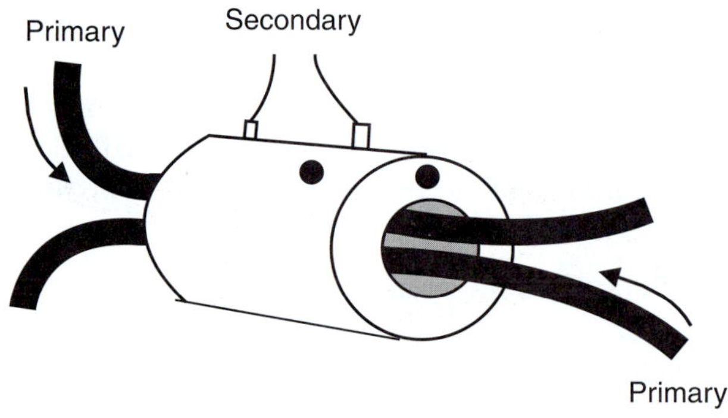

Figure 13–6   A 3-wire current transformer.

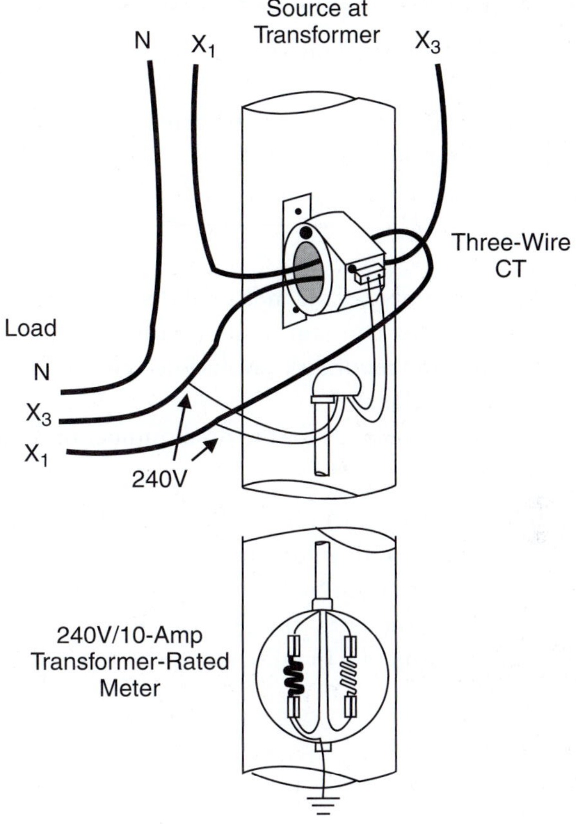

Figure 13–7   A 3-wire current transformer on a 120/240-volt service.

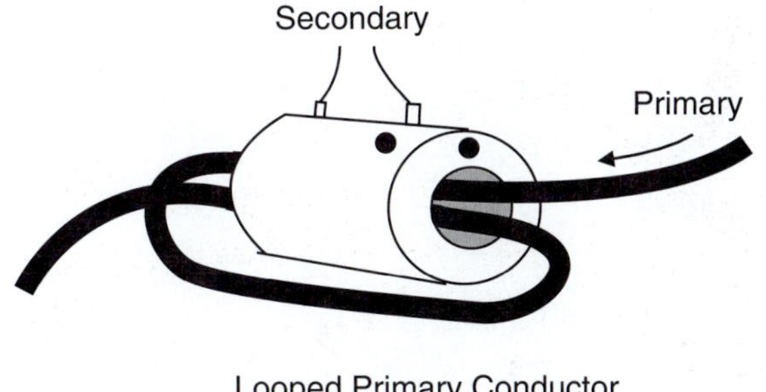

Figure 13–8   A double-looped primary.

13.4 Calculating Load, Using the Meter Spin Test

On occasion, it can be beneficial to know the actual instantaneous power demand used by a customer.

The instantaneous kW demand being registered by a self-contained meter can be determined by the following formula:

$$kW = \frac{3.6 \times Kh \times Rev.}{Sec.}$$

where   3.6 = the number of seconds per hour $\div$ 1,000

          Kh = the meter constant found on the meter nameplate

        Rev. = the number of disk revolutions (choose 10 for easier calculations)

        Sec. = time in seconds for the total number of revolutions counted

The meter multiplier (Kr constant) on a self-contained meter does not affect the spinning of the disk and is, therefore, not applicable to this formula.

To determine the load registered by a transformer-rated meter, multiply the answer in the preceding formula by the product of the current transformer (CT) ratio, voltage transformer (VT) ratio, and meter multiplier.

CHAPTER

14

Using the Truck and Boom

14.1 Check Out the Truck

14.1.1 Typical Daily Pretrip Inspection

A daily pretrip inspection is required by law for large trucks. If a truck has poor brakes, is overloaded, or has defective tires, any charges laid will be to the driver, not the mechanic or the utility/employer. Use the form provided and record the results in a logbook.

Prioritize the pretrip inspection on items that can lead to a catastrophic failure:

- Check for signs that indicate lug nuts are loose. An off-center valve stem in a truck wheel opening is an indicator of a loose wheel. Visual indicators, such as those shown in Figure 14–1, can also be installed.

- Too much travel (should be less than 2.5 inches/6 cm) in the slack adjusters of an air brake is a common fault leading to poor braking performance. The angle between the push rod and the adjuster arm should not be more than 90 degrees when the brakes are applied (see Figure 14–2).

Figure 14–1   Loose lug nut indicators.

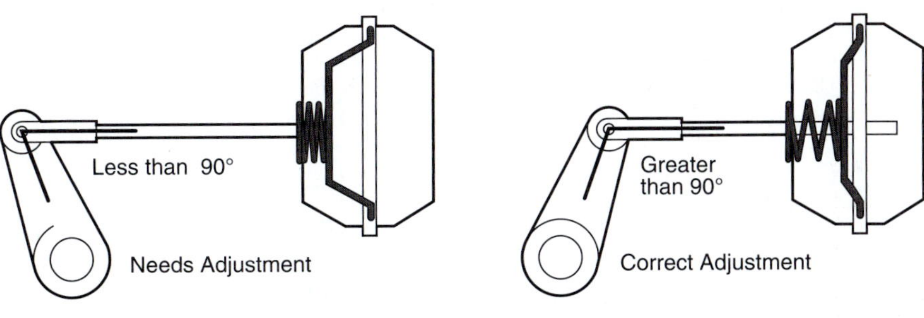

Figure 14–2 Slackadjusters with Brakes Applied.

- Check for a leaking hub or axle seal. If equipped with a sight glass on the wheel, check the oil level.

- Adjust trailer brakes so that they will not lock up when a brake pedal is applied. A trailer with locked-up brakes can slide sideways and strike other vehicles.

14.1.2 Summary of Pretrip Air-Brake Inspection

The following summary of a pretrip air-brake inspection is a memory jogger only and assumes that the driver has checked under the hood and will inspect the physical condition of the braking system and slack adjusters.

Step	Action	Details
1	Drain air tanks to zero pressure and then build up pressure again.	1. The low-air warning signal should cut out at 60 psi (415 kPa) minimum. 2. Pressure buildup between 85 to 100 psi (590 to 690 kPa) must be less than 2 minutes. 3. Governor cuts out between 100 and 135 psi (690 to 930 kPa).
2	At full pressure, fan the brake pedal to lower air pressure.	1. After a drop of about 25 psi (170 kPa) from the maximum pressure, the governor should cut in. 2. At about 60 psi (415 kPa) minimum, the low-air warning should cut in. 3. Between 45 and 20 psi (300 to 140 kPa), the parking brakes (spring brakes) should come on automatically.

(continued)

Step	Action	Details
3	Build back to full pressure.	1. With the engine stopped, fully apply the foot brake and hold for 1 minute. Pressure drop must be less than 2 psi (15 kPa). With a trailer, the drop must be less than 4 psi (30 kPa).
		2. With the engine at idle speed and the transmission in first gear, check the parking brake adjustment by trying to move forward.
		3. With the parking brakes off, drive ahead and apply the service brake as a check for response.
		4. With the truck in motion, activate the trailer brakes (spike) as a check for response.

14.1.3 Jump Starting a Vehicle

A battery gives off explosive hydrogen gas. A spark near a battery can trigger an explosion.

1. Before jump starting, check that the fluid level is normal and that the battery is not frozen. A maintenance-free battery should not be jump started if the indicator is red.

2. Only the last connection will cause a spark that could ignite the hydrogen gas. The last connection should, therefore, be a negative (ground) connection (Figure 14–3) at a location remote from the battery to the truck frame.

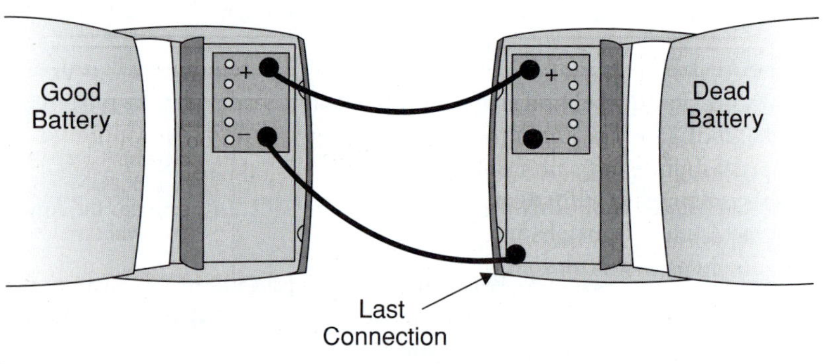

Figure 14–3 Jump starting a vehicle.

Studs are often installed on large trucks to facilitate jump starting. Look under the hood or near the driver's side step for a remote positive jump-starting stud, and look for a negative grounding stud on the truck frame. Turning on the headlights or some other load on the truck with the good battery truck will reduce the risk of a damaging voltage surge to electronic equipment.

14.1.4 Reduce High-Risk Driving Hazards

Lessons learned from utility truck accidents!

- Never use the hand valve to apply brakes only to the trailer. Locked brakes on a trailer can cause the trailer to slide sideways into other traffic.

- Do not jam on the brakes in an emergency. When the wheels lock up, the truck will go into an uncontrolled skid and steering will be impossible. For a truck without antilock brakes, use "stab" braking in an emergency stop. Emergency braking for a truck with antilock brakes requires steady and increasing pressure on the brake pedal. There may be variations for the most effective braking for trucks with antilock braking; check the owner's manual.

- Watch for restricted roadway height or width. Many derrick diggers and aerial devices are close to the maximum height of 13 feet, 6 inches (4.15 m) and width of 8 feet, 6 inches (2.6 m). A jib left out at an angle or a protruding outrigger can cause a vehicle to be over the allowable height.

- Be in the correct lower gear before starting down a steep hill. Overreliance on brakes to slow down a truck will cause the brakes to overheat and become ineffective.

- When extra space is needed to make a turn with a truck and trailer, take the extra space from the street you are entering, not from the street you are leaving.

14.1.5 Hauling Poles

Typical rules for hauling poles (each state may vary somewhat).

1. No long load permit is required if the combination truck, trailer, and load is less than 75 ft. (23m) and 5 ft. (1.5) or less overhang past the rear axle.

2. A long load (often annual) permit is typical for a truck trailer and load less than 92 ft. (28m) and 15 ft. (4.5 m) or less overhang past the rear axle.

3. Longer loads require a long load permit for each trip.

4. Long load permits are not typically valid at night or on holidays. For emergencies, notify the police and provide an escort.

5. A load of poles does not have any underride protection such as that found on tractor/trailers. Attach a flag, signs, or lights at night on any poles that overhang more than 5 ft. (1.5 m).

6. Make note of the maximum length of pole allowed on the pole trailer being used.

7. Make note of the maximum weight and, thereby, the number of poles allowed on the pole trailer being used. Tables 14–1, 14–2, 14–3, and 14–4 give estimates of some of the typical pole types hauled.

TABLE 14–1   Weight of Western Cedar Poles in Pounds

Length (ft.)	35	40	45	50	55	60	65	70	75	80	85	90	95	100
Class 1	1,050	1,320	1,590	1,790	2,030	2,290	2,820	3,170	3,700	4,400	4,840	5,810	6,750	7,500
Class 2	880	1,150	1,370	1,585	1,760	1,940	2,200	2,640	3,170	3,700	3,960	4,930	5,950	6,550
Class 4	660	790	1,010	1,230	1,410	1,670	1,940	2,290	2,640	3,080	NA	—	—	—
Class 5	570	700	880	1,150	1,410	NA	—	—	—	—	—	—	—	—

TABLE 14–2   Weight of Southern Pine Poles in Pounds

Length (ft.)	35	40	45	50	55	60	65	70	75	80	85	90	95	100
Class 1	1,570	1,880	2,220	2,590	3,000	3,450	4,020	4,620	5,200	6,400	7,200	8,140	6,940	NA
Class 2	1,350	1,620	1,910	2,210	2,570	2,940	3,340	3,780	4,330	5,170	5,750	6,400	6,000	NA
Class 4	1,000	1,220	1,440	1,690	1,930	2,190	2,460	2,730	3,020	3,615	NA	—	—	—
Class 5	860	1,060	1,270	1,500	1,720	1,950	2,240	2,490	NA	—	—	—	—	—

TABLE 14–3   Weight of Douglas Fir Poles in Pounds

Length (ft.)	40	45	50	55	60	65	70	80	90	100	110	120	125
Class H6	—	—	4,370	5,010	5,750	6,490	7,270	8,830	10,580	12,420	14,350	16,330	16,970
Class H5	—	—	3,910	4,550	5,150	5,840	6,530	7,960	9,520	11,180	12,880	14,670	15,640
Class H4		2,990	3,540	4,050	4,650	5,240	5,890	7,180	8,600	10,030	11,640	13,290	14,120
Class H3		2,710	3,170	3,680	4,190	4,740	5,240	6,490	7,730	9,060	10,490	11,960	12,700
Class H2	2,020	2,440	2,850	3,270	3,770	4,230	4,780	5,840	6,950	8,140	9,480	10,760	11,450
Class H1	1,840	2,160	2,570	2,990	3,400	3,820	4,280	5,240	6,300	7,360	8,510	9,710	10,350
Class 1	1,540	1,930	2,220	2,480	2,720	3,060	3,480	4,410	5,140	6,140	7,290	8,440	—
Class 2	1,310	1,610	1,870	2,130	2,430	2,750	3,080	3,960	4,740	5,680	6,660	—	—
Class 3	1,160	1,410	1,620	1,850	2,110	2,430	2,690	3,520	4,170	—	—	—	—
Class 4	1,160	1,410	1,620	1,850	2,110	—	—	—	—	—	—	—	—

TABLE 14–4   Sample Weight of Concrete Poles

Pole Length (ft.)	Pole Class	Weight (lbs.)
30	B	1,390
35	A	1,600
35	B	1,750
35	E	2,250
40	E	2,750
45	E	3,230
45	F	3,560
45	H	4,260
50	G	5,000

14.2 Monitoring a Hydraulic System

14.2.1   Troubleshooting a Hydraulic System

If	Then
There is any suspicion of a hydraulic fluid leak and its location is unknown.	Do not try to find the leak by touching any hydraulic lines while the system is under pressure. Hydraulic fluid coming from a pinhole in a hydraulic hose under pressure can pierce the skin, which can be a serious injury/infection and needs medical treatment.
A high-pitched whine is heard.	Warm up the fluid when the temperature is below freezing. A typical warm-up would be to set the engine speed at 600 rpm (1,000 rpm for diesel engines), engage the hydraulic pump, let it run for about 10 minutes, and then increase the engine speed gradually until it reaches the specified rpm. **Caution:** The engine rpm may increase on its own as the unit warms up, and this could over-rev the pump.
A sound like the hydraulic pump is pumping marbles is heard (cavitation).	A partial vacuum in the fluid at the pump is the cause of this cavitation and eventual hydraulic pump failure. Cavitation occurs when the pump is over-revved, brought up to speed before the fluid has warmed up, the fluid level is low, or the fluid filter or a suction line is restricted.

(continued)

14.2.1   Troubleshooting a Hydraulic System *(continued)*

If	Then
A loud shotlike sound (*water hammer*) is heard.	This is caused by a sudden stoppage of the fluid in the system. A resulting pressure surge can be as high as four times the normal pressure. Feathering the controls (easing the valve open slowly) or some corrective plumbing may be needed to prevent this. Valves controlled by radio or by fiber optics are designed to prevent shock loading.
The cylinders have jerky or erratic movement.	Air may be in the system, or a cylinder rod may be bent.
The hydraulic pump or a hose is too hot to touch.	Oxidation in the fluid will cause sludge to form and create more heat and an eventual breakdown.
A high-frequency vibration can be felt when touching steel fittings.	This is a sign of a damaging condition that needs correction.
The oil has a milky appearance.	The oil is saturated with air or water.

14.2.2 A Conductive Vacuum in a Hydraulic Hose

The weight of fluid in a hydraulic line having a separation distance of more than 35 feet (11 m) between the oil reservoir and the upper end of the hydraulic line will drop in the line somewhat and drain into the reservoir. The partial vacuum formed at the top of the hydraulic line has a lower resistance to an electrical flashover than does hydraulic fluid or air at atmospheric pressure.

Check valves installed in the hydraulic line to prevent the fluid level in the line from dropping and creating an electrically conductive vacuum. Vent valves are installed, as a backup, to allow air to enter the hydraulic line and to keep it at atmospheric pressure. The filter in the valve keeps dirt out. These valves need cleaning. These valves are critical for barehand work where the insulation of the boom and hydraulics provide primary protection.

14.2.3 High-Pressure Hoses

Compression tools, wrenches, hot-line tools, and tree-trimming tools being operated from the tool outlet at the bucket level must use hydraulic hoses rated for 10,000 psi and must be nonconducting (and are, generally, orange). A metal-reinforced hydraulic hose, used by mistake, has caused an electric short that resulted in a ruptured hose and a hydraulic fluid fire in the bucket.

14.2.4 Check Holding Valves

A holding valve will keep a boom up in the air even if a hydraulic hose fails. The following describes how to do a drift check on the holding valves:

1. Extend the outriggers.

2. Put the boom into the positions shown in Figures 14–4 and 14–5.

3. Shut off the vehicle.

4. Activate the valves to lower outriggers and booms in the various positions shown in Figures 14–4 and 14–5. Hold the positions for 15 seconds.

Any movement downward indicates that the holding valve needs maintenance or that air is in the system. A formal drift test can be done in the same way, except that the boom is loaded with its rated load and held for a given period of time.

14.2.5 Emergency Bucket Lowering

Use the lower controls to bring down the boom in an emergency. The lower controls override the upper controls. It is a given that the crew member on the

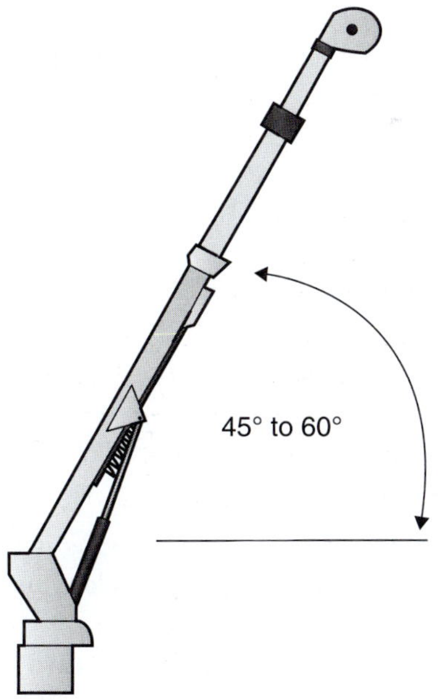

Figure 14–4 Position of a digger derrick for a drift test.

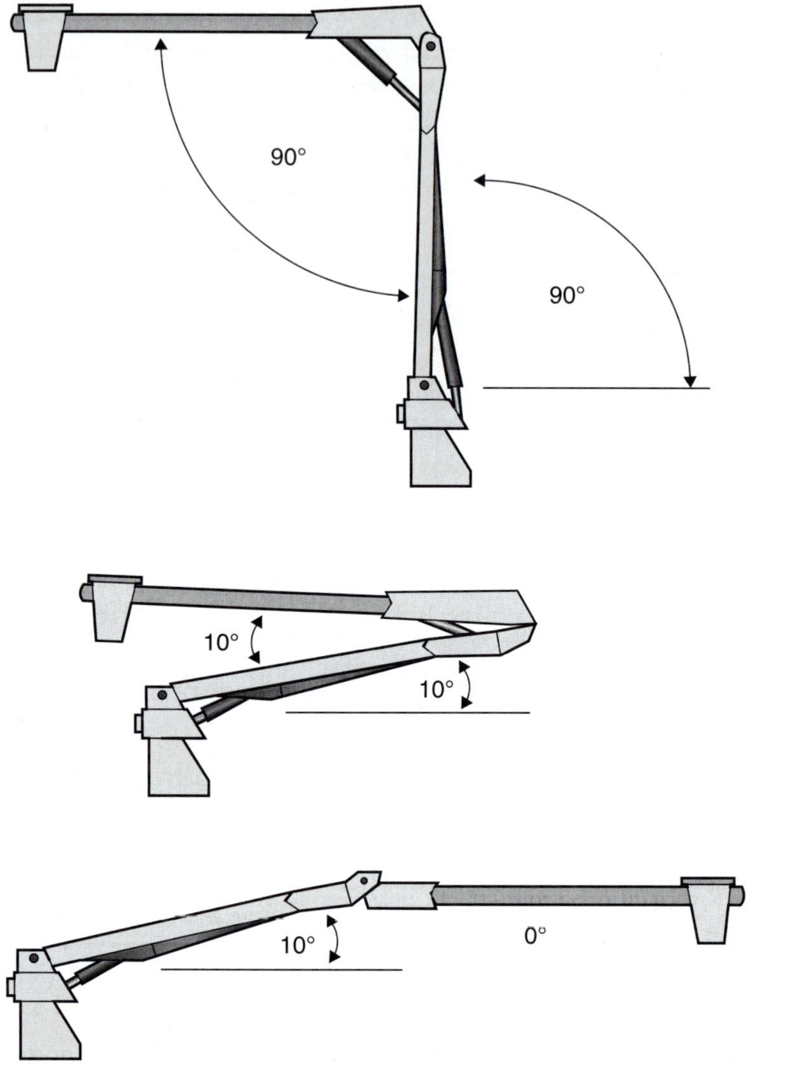

Figure 14–5 Position of an aerial device boom for a drift test.

ground has been trained and has practiced bringing down the boom using the lower controls, removing a casualty from the bucket, and performing necessary CPR and first aid.

When hydraulic power is lost and the boom has to be lowered, some units have an auxiliary battery-operated pump to lower the boom and others have hydraulic couplers that allow the hydraulic system of another truck (preferably a truck with clean insulating fluid to prevent contamination) to connect and operate the disabled system.

If the boom has to be lowered by manipulating the holding valves, anyone in the bucket(s) should first lower themselves from the bucket to the ground using an approved, and practiced, method and equipment. Once no one is in the bucket, there remains no urgency to lower the boom and store the outriggers. Generally, no-power lowering is best left to a mechanic or specialist because it is very easy to lose control of the process.

14.3 Stabilizing a Boom-Equipped Vehicle

14.3.1 Procedure for Stabilizing a Boom-Equipped Vehicle

Step	Action	Details
1	Ensure that the mechanical and hydraulic systems of the unit have been maintained.	Ensure that normal maintenance has been carried out. 1. Learn the location and check critical welds for surface cracks and lines of rust. 2. Check pin retainers at pivot points. After any repair or hydraulic oil leak, put the boom through its full cycle to ensure that all hoses are full of oil and that there will be no sudden collapse of the boom because of an air pocket. On a scheduled basis, do a drift test on the holding valves. Keep a log book on the hydraulic unit.
2	Apply the parking brake and chock blocks.	Hydraulic-braked trucks with boom equipment have both a mechanical parking brake and a hydraulic locking brake (*accumulock*). Both must be on. **Caution:** Hydraulic parking brakes can bleed off. Should this happen when the job is finished and the outriggers are raised, the truck will be free to roll down a slope unless chock blocks and the mechanical parking brake are applied. If they are properly adjusted, the parking brakes on air-braked trucks are very reliable.

(continued)

14.3.1 Procedure for Stabilizing a Boom-Equipped Vehicle *(continued)*

Step	Action	Details
		With the vehicle on an incline, extending the outriggers reduces the holding power of the tires. Use wheel chocks to help prevent the vehicle from sliding downhill.
		Actions of the boom sometimes cause the truck to shift on the outrigger planks, especially if the wheels are off the ground. Stability can be increased by anchoring the truck to another truck and/or installing chock blocks, front and back.
3	Use outrigger pads and other means to provide a solid surface for outriggers.	Loss of stability can occur because the outriggers penetrate into the earth, asphalt, or concrete or because the vehicle slips off the outrigger pads. Each outrigger should be on pads to offer a greater surface-for-weight distribution.
4	Extend outriggers (stabilizers).	The outriggers should lift all the weight normally supported by the springs. Some telescoping outriggers will have a mark indicating the minimum extension needed to provide stability.
		Units with only two outriggers (trucks with the turret mounted just behind the cab) will sometimes have the front wheels lifted from the ground. Planking or support under the front wheels is needed to prevent a sudden shock-loading drop when the boom is rotated toward the front of the truck.
		Keep the outrigger in sight, or make sure that all people are clear while lowering the outrigger.
5	Level the vehicle to less than 5° on a slope.	On sloping ground, extend the low-side outrigger first, then extend the other outrigger(s) until the unit is level. It may be necessary to build up the low side with extra timbers or to dig out the high side or to use the pads specifically made for slopes. If a telescoping-type outrigger cannot be extended far enough, it creates a stability hazard. A radial-type outrigger has an advantage over a telescoping-type outrigger for this reason.

(continued)

14.3.1 Procedure for Stabilizing a Boom-Equipped Vehicle *(continued)*

Step	Action	Details
6	Trucks without Outriggers	The truck must be set up as level as possible. A boom over the side can put a truck without outriggers on a 3-degree tilt. If the truck is set up initially on a 5-degree slope, the unit could be operating on an 8-degree tilt.
		The low side can be built up by driving up on pads or similar material.
		It is critical to the stability of the truck to maintain proper tire pressure and to learn the inspection points for the stability features of the vehicle.

14.4 Electrical Protection for Working with Truck Booms

14.4.1 Electrical Protection Around Trucks and Booms

The only real protections available when a boom makes contact with a live conductor are as follows:

1. *Staying on the vehicle or on a ground-gradient mat* bonded to the vehicle will keep a person at the same potential as the vehicle. As long as a person avoids contact with anything not connected to the vehicle, no current will flow through that person to another object.

2. *Keeping a safe distance away from a vehicle* will keep a person from making inadvertent contact and will keep a person away from any high ground.

14.4.2 Minimum Approach Distance for Truck Booms

In most jurisdictions, occupational safety regulations do not allow an unqualified operator to operate a boom within 10 feet (3 m) of a distribution electrical circuit and an additional 4 inches (10 cm) for every 10 kilovolts over that, which works out to 14 feet (4.3 m) for 169 kilovolts, 16 feet (4.9 m) for 230 kilovolts, and 25 feet (7.6 m) for 500 kilovolts.

A qualified operator may operate a boom up to the minimum approach distances shown in Table 2-1. The truck would be grounded and a dedicated observer used where visibility is questionable. The boom is allowed to approach and leave a space of at least 6 inches (15 cm) to distribution conductors that have been covered up.

14.4.3 Types of Contact Hazards with Noninsulated Booms

If	Then
A person, on the ground, is *in contact* with a utility vehicle, winch line, or power-installed anchor while the noninsulated portion of a boom contacts a live circuit.	The person is a parallel path to ground and will take *some* share of the current flowing to ground.
A person on the ground is *near* a utility vehicle while the noninsulated portion of a boom accidentally contacts a live circuit.	The person is exposed to ground gradients at each location where current is entering the earth.
In a high-induction area, a noninsulated boom is used to lift or handle an *isolated and grounded* conductor.	Unless the boom is bonded to the grounded conductors, the boom can be at a different potential than the conductors being handled.

14.4.4 Electrical Protection Working with a Noninsulated Boom

Action	Details
Reduce the risk of a boom contact with a live conductor.	1. *Maintain minimum approach limits.* Use a signal person (or radio with crane) to watch the approach distance to exposed lines and equipment, as illustrated in Figure 14–6. 2. *Use cover up where applicable.*

(continued)

14.4.4 Electrical Protection Working with a Noninsulated Boom
(continued)

Figure 14–6 Hand signals for derrick operations.

(continued)

14.4.4 Electrical Protection Working with a Noninsulated Boom
(continued)

Action	Details
People on the ground must stay away from the vehicle trailer.	The operator of a boom should ensure that everyone stays away from the vehicle when the boom is to be moved near a live circuit. One option is barricading. However, if the barricades are put in place before they are needed, workers tend to routinely violate the barricades as they prepare a transformer or frame a pole.
Ground the truck to promote a fast trip-out of the circuit if the truck should become alive.	A fast trip-out reduces risk by reducing the length of exposure to the hazard; it does not eliminate a potentially fatal shock hazard for anyone touching the vehicle. The following are typical ground electrodes, listed in priority: 1. A permanent ground network such as a station ground, a neutral, or a tower ground 2. An existing ground rod or an anchor rod in earth 3. A temporary driven ground rod
Bond the protective grounds on an isolated line to a truck ground to eliminate a voltage difference between a circuit (especially in a high induction area) and a remote ground or truck boom.	If a noninsulated boom is used to handle conductors on a job where a circuit is isolated and grounded, the boom should be bonded to the protective line grounds. On a transmission right-of-way that has high induction from live neighboring circuits, or in the rare case of accidental reenergization, bonding ensures that there will be no potential difference between the vehicle, the boom, and its wire rope winch and conductors.

14.4.5 Electrical Protection Working with an Insulated Boom

Action	Details
Maintain a minimum approach distance.	When a qualified operator is in the bucket of an electrically tested bucket truck, that bucket and boom should maintain the same minimum approach distance as a qualified worker must maintain when working on a hot line. Maintaining this distance will reduce the risk of inadvertent contact with other phases or objects at different potentials while working with rubber gloves or barehanded.
Apply cover-up on the conductors.	After protective cover-up is installed, a visible air space should be kept between the insulated portion of the boom and the covered conductors.
Monitor the electrical clearance from the lower boom.	There have been instances of the noninsulated lower boom of an aerial device contacting a live underbuild or a lateral tap. These vehicles must be grounded to protect workers on the ground. The need to ground an aerial device with an insulated lower boom insert is based on individual utility preference. **Caution:** Some lower-boom inserts may have had their insulation short circuited to use the contamination monitoring system. **Caution:** If a unit with an insulated lower-boom insert (which has not been shorted out) is used for work on a transmission line, a high voltage can be induced on the section of boom above the insert.
Maintain the electrical integrity of the boom.	See section 14.4.6.
Monitor for boom contamination and leakage.	See section 14.4.7.

14.4.6 Taking Care of the Insulated Boom, Buckets, and Jib

- Keep the insulated section of the boom, bucket, and jib clean and dry. Maintain a waxed or silicone surface to repel water. Use specified cleaners only; abrasive cleaners can leave tiny scratches, and solvents can soften the surface coatings. A dry boom with a chalky or nonwater-repelling surface may pass the dry dielectric test and still fail electrically after a brief shower.

- Keep the boom interior clean and dry. Wash it out with a low-pressure washer (clean water only), rinse thoroughly, and dry it by leaving the boom in a vertical position or by pouring isopropanol down the boom as a drying agent. High-pressure water can cause water to diffuse through the fiberglass, which will require a very long time to dry out.

- Ensure that insulated booms, buckets, liners, and jibs get their scheduled electrical retests. A boom dielectrically tested while it is still wet can cause permanent damage.

- Boom and bucket covers prevent road salt and road wash from contaminating the insulated portions of a unit. They can also prolong the life of a boom by protecting it from the ultraviolet rays of the sun, which can cause the boom to look fuzzy when the fibers become exposed.

- Cover the jib or store it in a canvas bag or secured on padded holders in a truck bin.

- Boom leakage over 1 milliampere (1,000 microamperes) can leave track burns, typically inside the boom. If when checking out the boom for barehand work or carrying out a formal electric test the reading on the meter shows a return current approaching 1,000 microamperes, stop, clean, or dry the boom.

- Strap down the boom during travel. Vibration and shock loading will shorten the life of fiberglass, and the boom will be damaged where it sits on the boom rest.

14.4.7 Monitor Boom Leakage

Barehand work and high-voltage rubber-glove work typically require that the aerial device have a boom-contamination monitoring circuit that will measure and monitor the actual leakage current of the insulated boom. The initial certifying electrical proof test should be used as a reference point for future contamination monitoring. Keeping a log of every test will show trends that may develop. The monitoring circuit of the aerial device (Figure 14–7) is checked before doing the boom-contamination test. The circuit can be tested for an open or short circuit using the "test" position on the same meter used to measure the leakage current.

A current test is made daily before starting work and before working on a higher voltage. A metal part of the boom or bucket grid is put in contact with the energized circuit to be worked on at 1-minute intervals.

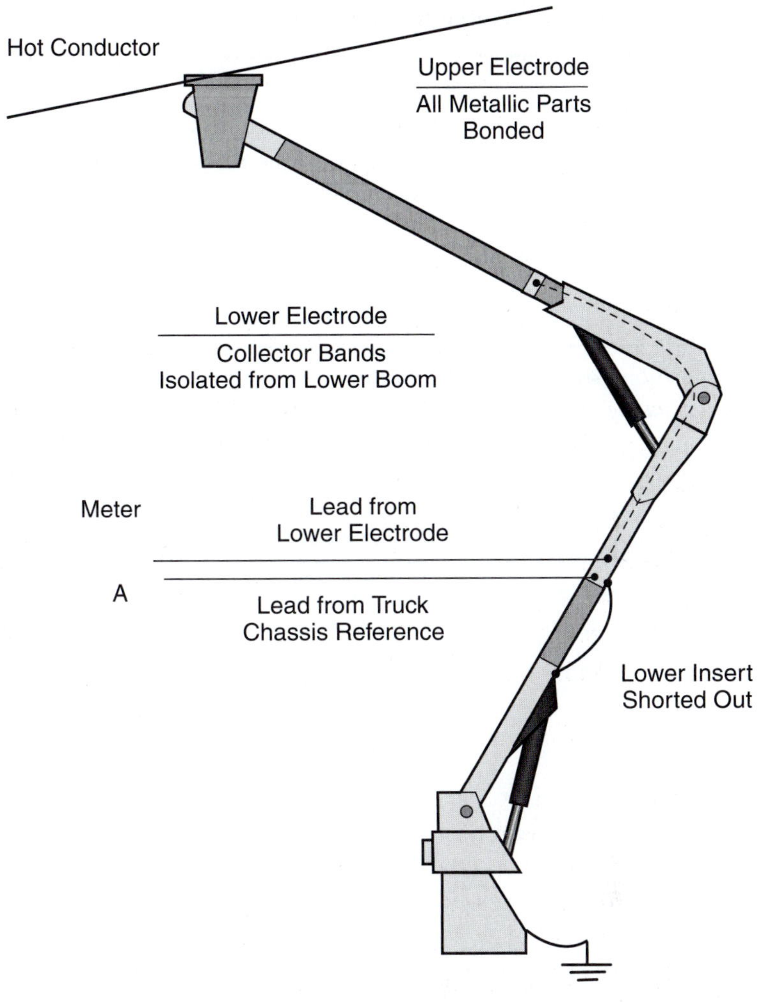

Figure 14–7 A boom-contaminating monitoring circuit.

The microampere meter shows the amount of current between the upper electrode, which consists of all the metal components bonded together, and the lower electrode, which is the collector band mounted inside and outside of the boom, isolated from the metal lower boom. The leakage current should not exceed 1 microampere per kilovolt of the phase-to-phase voltage of the circuit. Many units will be well below this standard.

Caution: Intentionally letting a boom with high leakage current stay in contact with the circuit as a means to dry the boom will leave carbon tracking inside the boom, which will eventually lead to permanent damage and an electrical failure.

14.4.8 Contamination Meter Readings

If	Then
The contamination meter reading is *high* (over 90) or there is a *sudden increase* in the contamination meter reading.	1. Try retesting with the lower-boom position farther away from the conductor (lessen induction). 2. If not successful, clean, dry, and retest the boom. 3. If not successful, remove truck from service and get it checked by specialists.
The contamination meter reading is *fluctuating* up and down.	1. Moisture is probably present. Wipe down the boom and remove the moisture inside of the boom by pouring isopropanol into it. 2. If not successful, there may be a fault in the boom insulation. Remove the truck from service and get it checked by specialists.
The contamination meter reading is showing a *gradual increase* from day to day.	1. There may be dirt contamination on the inside and outside surfaces of the boom. Clean the boom inside and out, let the boom dry out, and then retest. 2. The hydraulic oil may have become contaminated. Shut down the unit for about 15 minutes, drain the water from the reservoir, and then retest. If not successful, change the hydraulic oil and retest. 3. If not successful, there may be a fault in the boom insulation. Remove the truck from service and get it checked by specialists.

14.5 Operating a Digger Derrick

14.5.1 Lifting with a Digger Derrick

Reduce the risk of damaging a digger derrick boom by using the lifting charts and boom features available.

1. The lifting capacity chart for each type of digger derrick must be available to the vehicle operator. A lifting chart similar to the one shown in Figure 14–8 should be visible from the boom operator's position.

2. Lifting capacity is decreased as the boom is extended away from the turret location (see Table 14–5).

3. The winch and stinger are strong enough to lift loads in a position that will overload the boom.

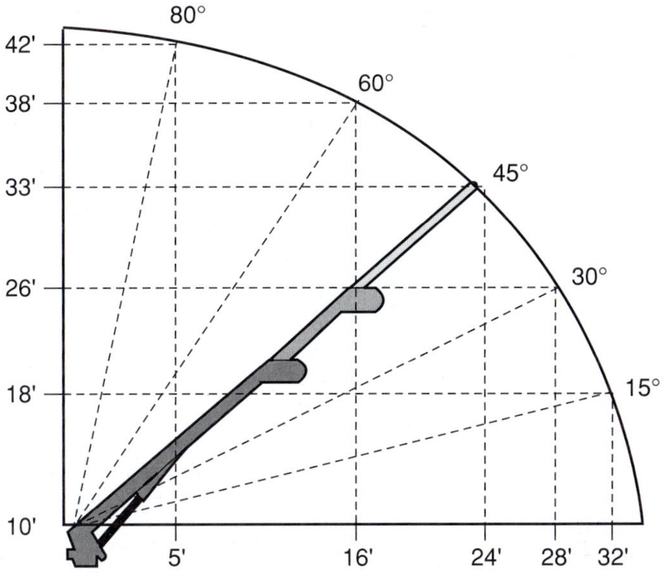

Figure 14–8   A typical digger derrick lifting-capacity chart.

TABLE 14–5   Lifting Capacities for Digger Derricks

All Booms Retracted	Boom Angle	80°	75°	60°	45°	30°	15°	0°
	Lbs. Load	9,500	7,900	5,500	4,300	3,700	3,350	2,950
Intermediate Boom Extended	Elevation	80°	75°	60°	45°	30°	15°	0°
	Lbs. Load	7,000	5,900	3,600	2,800	2,500	2,100	1,850
3rd-Section Capacity	Elevation	80°	75°	60°	45°	30°	15°	0°
	Lbs. Load	4,000	3,400	2,000	1,450	1,200	1,100	1,000

4. The lifting capacity changes with the direction of the boom in relation to the truck, outriggers, and slope of the ground. Figure 14–9 provides a sample of the derated lifting capacity of a corner-mount two-outrigger unit on an uphill slope.

Note: If the unit is equipped with overload protection, certain boom functions will not operate when the unit is overloaded. To check the overload protection, raise the boom to its maximum height until the system bypasses and then try to extend the stinger. The stinger will not extend if the overload protection is working. Lowering the boom to the bottom will reset the overload protection.

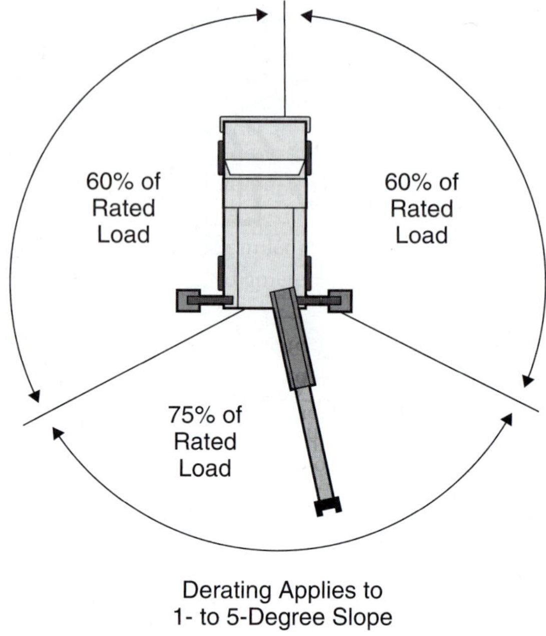

Figure 14–9   Load capacity derated on slope.

14.5.2 Rotation Gear Failure Hazard

A rotation gear failure allows the boom and load to rotate freely into objects and people. The rotation gears are often the weakest link when pulling an object from a ditch, operating a heavy load with the truck set up on a slope, or installing a power anchor.

- Do not pull a load as shown in Figure 14–10. Rotate the boom toward the load to make the lift.

- Set up so that a power-installed screw anchor can be installed using very little rotation. If rotation is necessary, make sure the boom is moved to follow the anchor as it is installed.

- Do not try to loosen a pole setting for removal by rotating the boom side to side.

- Derate the lifting capacity of the boom when set up on a slope.

14.5.3 Lifting with a Digger Derrick Winch

The lifting capacity of a boom tip or turret winch and the strength of the winch cable or winch rope can be greater or less than the capability of the boom or load.

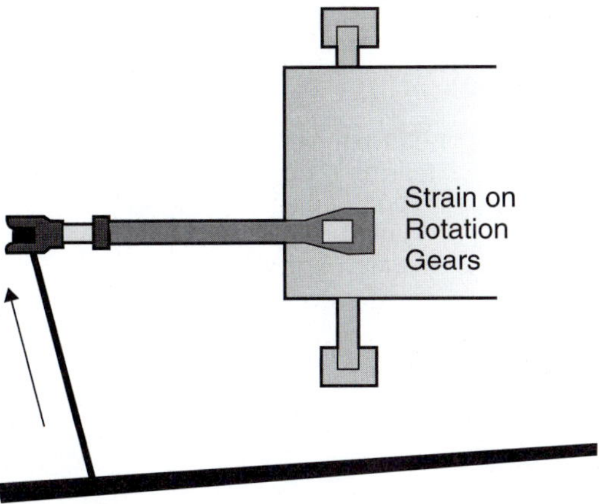

Figure 14–10   Rotating a load.

When the Winch Is Too Strong

A 1/2-inch wire rope winch cable (Table 14–6) with a 4,200-pound maximum working load is strong enough to break a boom or tip over a truck while lifting with the boom in one of many weak positions (seen in Figure 14–9 where a truck lifting capacity is derated on a slope). Make all lifts by booming up first, when the load is suspended, then lift with the winch.

When the Winch Is Too Weak

A digger derrick, with the boom in a vertical or strong position, will lift beyond the maximum working load of a 1/2-inch winch cable. A two-part line must be used to lift heavier weights than the maximum working load of the winch. Figure 14–11 shows the decreased tension on the cable when using a two-part pull.

TABLE 14–6   Typical Digger Derrick Winch Cable Capacities

Maximum Working Load for Winch Cable/Rope	
0.5-inch double-braided polyester rope	2,000 lbs./910 kg
0.75-inch double-braided polyester rope	3,500 lbs./1,590 kg
1-inch double-braided polyester rope	6,000 lbs./2,700 kg
1.25-inch double-braided polyester rope	10,000 lbs./4,500 kg
0.5-inch wire rope	4,200 lbs./1,900 kg

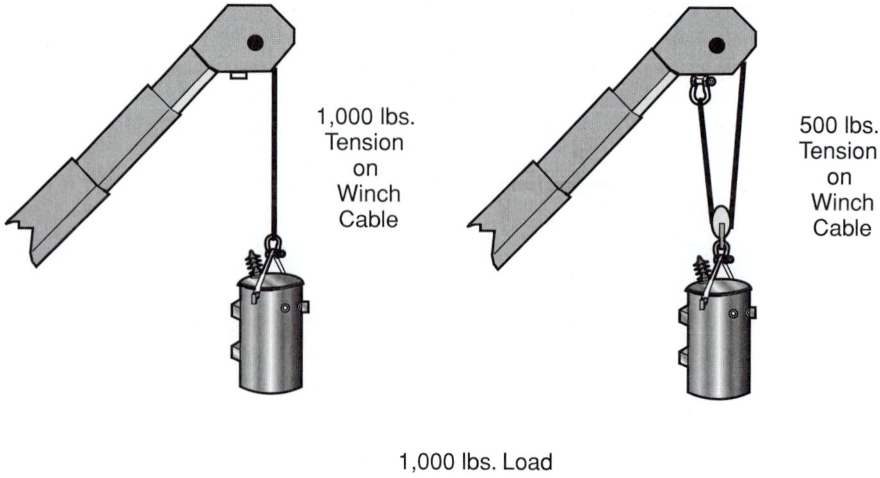

Figure 14–11 A boom-tip switch.

Winching Failures

- If the winch creeps down under load, the winch brake needs maintenance.

- A rotating load on a winch will damage the winch cable and cause a premature failure. Use a tag line and/or a swivel-type load hook.

14.5.4 Pulling a Pole

To remove a pole from the ground, use a hydraulic pole jack (butt puller). If necessary, use the auger to loosen a pole setting, then lift with the boom only. Do not use the winch, the stinger, or the boom rotation to loosen the pole in the ground. Do not use the auger as a stiff leg.

14.5.5 Installing a Screw Anchor

The torque needed to install a screw anchor varies with soil conditions, and it is, therefore, not difficult to apply more torque than the anchor can withstand. A unit with a "torque limiter" control can be set to limit the torque generated by the auger motor. Screw anchors have a torque rating stamped on them. When installing a screw anchor, the anchor will draw the boom into the direction the anchor is going into the ground. This has been the cause of boom and rotation gear failures. Operate the boom so that it follows the anchor as it is screwed into the ground. Reduce the complexity of following the anchor by setting up the unit so that it is not necessary to rotate the boom, as well as by booming down. The risk of boom failure is reduced when the overload protection is working properly.

14.6 Operating an Aerial Device

14.6.1 Aerial-Device Boom Lifting Capacity

Variables	Effect on Boom Capacity and Stability
The lifting capacity is dependent on stability.	The stability of the truck setup affects the ability to lift the total weight shown on the lifting charts. Lifting charts assume a level, stable setup. Some vehicles will have a derating factor for vehicles not parked on the level.
The lifting capacity is dependent on boom angles.	Figure 14–12 provides *a sample only* of the information about lifting capacity that must be taken into account when operating an aerial device. Always use a lifting chart or table for the specific boom being used.

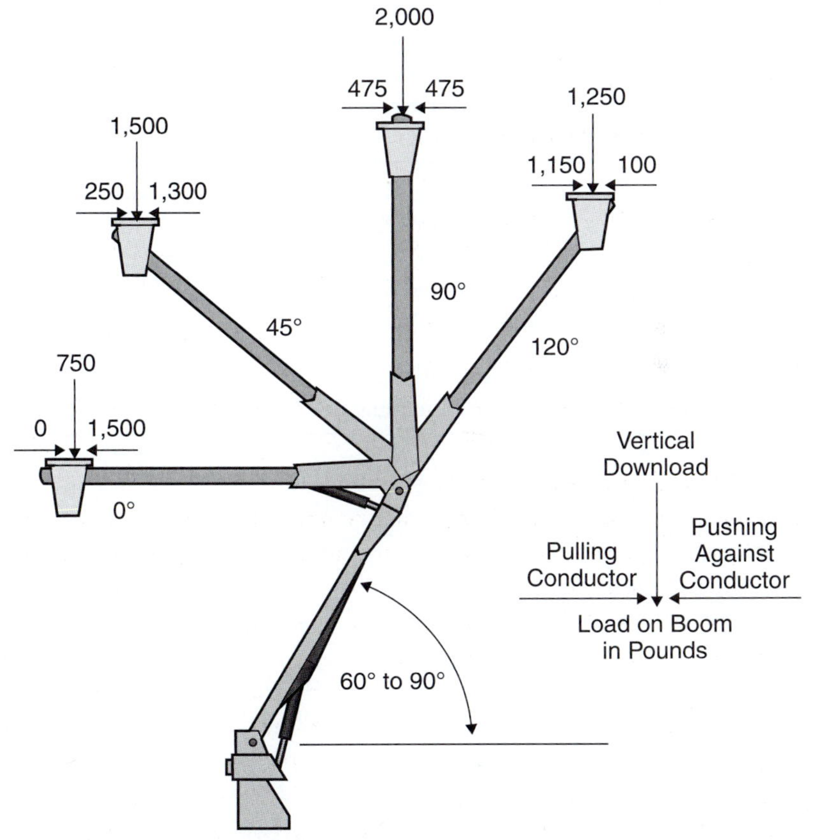

Figure 14–12 A sample lifting-capacity chart.

(continued)

14.6.1 Aerial-Device Boom Lifting Capacity *(continued)*

Variables	Effect on Boom Capacity and Stability
	Caution: Determine whether or not the capacity chart you are using refers to "bare shaft capacity." If so, the total weight of the buckets, jib, and workers must be subtracted from the capacities shown on the chart.
The lifting capacity is dependent on horizontal loads.	An aerial device is often used to lift conductors into and out of corner structures. Many manufacturers will state that no horizontal load is to be placed on their units. Horizontal loading-capacity charts are not available for many units. Moving conductors into a corner is a high-risk operation because the magnitude of the load is usually unknown and the load increases quickly as the conductor is moved into the bisect. The horizontal loads shown in Figure 14–12 are sample numbers only. **Note:** Some boom configurations have virtually *no* horizontal load capacity. There is virtually no capacity to pull a conductor at a corner.
Jib capacity	The loading of a jib is also dependent on the angle and length of extension (Figure 14–13). While the boom may be at a good angle to make a maximum lift, the jib angle and extension may very well be the weakest link. **Caution:** The winch and hydraulic jib used on material-handling units are for vertical loads only.

14.6.1 Aerial-Device Boom Lifting Capacity *(continued)*

Variables	Effect on Boom Capacity and Stability

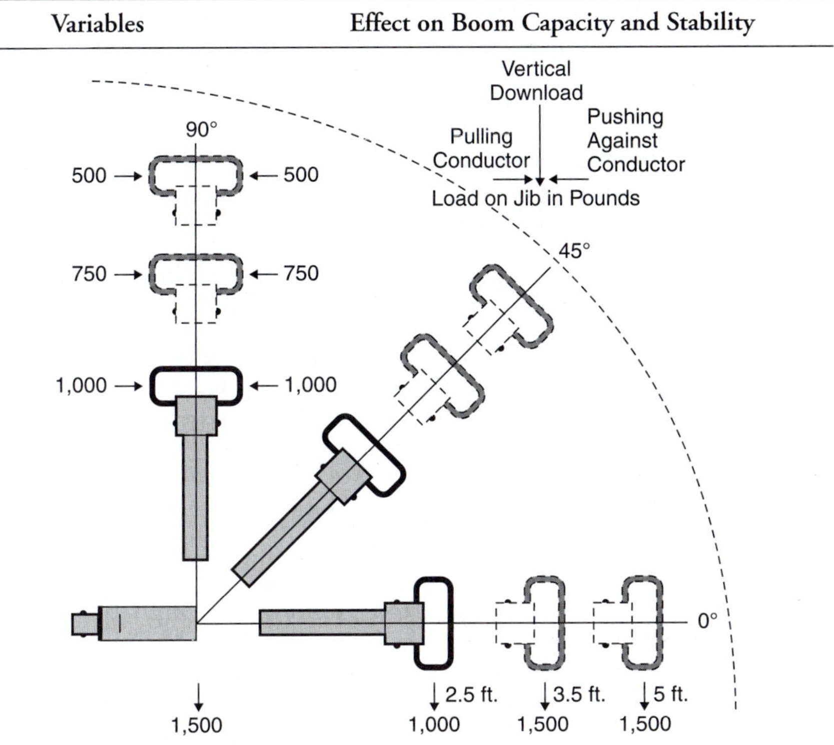

Figure 14–13 A sample jib-capacity chart.

14.6.2 Working with a Material-Handling Aerial Device

The risks involved with operating a material-handling aerial device must be reduced, so pay attention to the following:

1. Keep the load low, and raise or lower it only after reaching the position where the load is to be installed or removed.

2. Remember that the *three lifting variables* are known only after reaching the position where the load is to be installed or removed:

   - Upper-boom angle (Check the angle indicator on the boom.)
   - Lower-boom angle (Check the angle indicator on the boom.)
   - Load-line radius (How far is the jib extended out from its mounting?)

3. Check the capacity chart for the net load the unit is able to lift in the position. The chart should be easily visible to the operator.

4. Side loading is hazardous. Make vertical lifts only.

5. Maintain your minimum approach distance to live conductors when the winch is in contact with the ground, the pole, etc. A rope winch line eventually becomes conductive because of moisture and contamination.

6. Use a dynamometer to determine the weight of an unknown load.

7. Watch for "winch pileup." When the rope falls off the buildup on one edge of a winch, a shock load can be severe enough to cause a rope to fail, a transformer to unload, or an outrigger to sink into the ground and cause a truck to tip over.

8. The capacity of a winch is specified at its first layer of rope. A buildup of rope will derate the winch capacity significantly.

Rigging in Powerline Work

15.1 Working Load Limits

The term *working load limit* (WLL) is used for rigging in this text because a term like *safety factor* can be misleading. The WLL for individual pieces of equipment does not necessarily mean that a rigging setup configuration is a safe working load. The WLL of each piece of rigging hardware must be labeled or identified by size (diameter) and checked against WLL tables. When using rigging equipment such as rope blocks, hand line, fiber rope, wire rope, nylon slings, blocks, transformer davits (gins), snatch blocks, anchor pulling eyes, conductor grips, and chain hoists, the WLL must be known.

15.2 Rigging Hardware

15.2.1 Fiber Rope

For accurate strength ratings, consult tables for the WLL of the fiber rope to be used. When accuracy is not critical, the following rule of thumb can used as an indicator of the WLL:

Rule of Thumb for Fiber Rope	Example
1. Convert the rope diameter to eighths.	To find the WLL of 1/2-inch polypropylene rope:
2. Square the numerator.	
3. Multiply by:	1. Translate 1/2-inch diameter to eighths of an inch, which is 4/8.
• 40 for three-strand polypropylene rope	2. Square the numerator 4^2: $=16$.
• 50 for hollow-braid polypropylene	3. Multiply 16 by the rule-of-thumb factor of 40.
• 50 forpoly-dacron rope	4. WLL is, therefore, $40 \times 16 = 640$ pounds.
• 70 for three-strand nylon rope	
• 90 for double-braid nylon rope	
4. = approximate WLL in pounds.	

Derating factors for the WWL of fiber ropes when using knots and splices.

- 10 percent for an eye splice
- 45 percent for a bowline knot
- 35 percent for a running bowline knot
- 60 percent for a square (reef) knot
- 50 percent for a round turn and two half hitches

15.2.2 Wire Rope

For accurate strength ratings, consult tables for the WLL of wire rope to be used. When accuracy is not critical, the following rule of thumb will be an indicator of the WLL:

Rule of Thumb for Wire Rope	Example
1. Convert the rope diameter to eighths.	To find the WLL of 5/8-inch regular-laid wire rope:
2. Square the numerator.	
3. Multiply by:	1. 5/8 diameter, has a denominator of 8.
• 250 for regular-laid wire rope (6 × 19 and 6 × 25)	2. Square the numerator 5^2: = 25.
• 150 for cable-laid wire rope (3 × 3 × 19).	3. Multiply 25 by the rule-of-thumb factor of 250.
4. = approximate WLL in pounds.	4. WLL is, therefore, 250 × 25 = 6,250 pounds.

Making a Temporary Eye

Figure 15–1 shows the correct way to use U-bolts to put a temporary eye in a wire rope. The clips are spaced at a distance equal to six times the diameter of the wire. The clips should be tightened again after strain is put on the rope and also should be checked regularly. The clip farthest away from the eye gets most of the vibration and is usually the first to loosen.

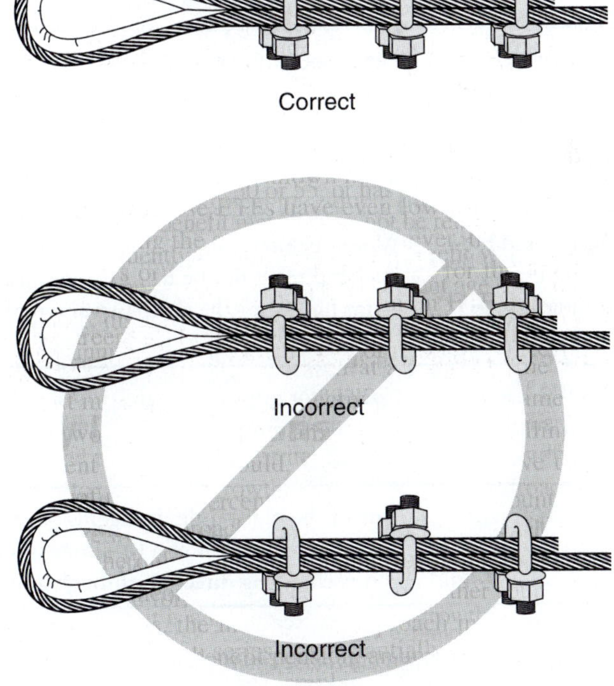

Figure 15–1   Using U-bolts to put a temporary eye in a wire rope.

Specific Hazards when Working with Wire Rope

Hazards	Barriers
Sharp bends in a wire rope sling will derate the strength of the sling.	Provide padding, such as wood blocking, when a sling is used around sharp edges. As a rule of thumb, the diameter (D) of the rope bend in relation to the diameter (d) of the rope itself is as follows: 1. $D/d = 10$. No reduction in strength: for example, a 1-inch cable bent around a 10-inch object. 2. $D/d = 2$. The sling is derated to 65 percent of its original strength: for example, a 1-inch cable bent around a 5-inch object. 3. $D/d = 1$. The sling is derated to 50 percent of its original strength: for example, a 1-inch cable bent around a 1-inch object.
An eye in wire rope will derate the strength of the rope.	Wire-rope strength is reduced 20 percent for a U-bolt-clipped eye. Wire-rope strength is reduced 10 percent for a Flemish eye.
Guy steel used as a winch will twist to release a wire-form grip (preform), and it will break if used through a snatch block.	Never substitute guy steel for wire rope.
Wire rope typically used for stringing will break if made into slings and subjected to sharp bends.	Wire rope used for stringing is made of harder steel and must not be made into slings. This steel is not meant for the sharp bends typical for slings.
Conductor grips (Chicago style) are easily overloaded when used on wire rope.	Conductor grips (Chicago style) should not be used on wire rope.
When standing in the bight of a wire rope under tension, any rigging failure can be fatal.	One of the oldest rules in rigging states to always stay out of the bight of a wire rope under tension as it is changing direction through a block.

15.2.3 Chains

For accurate strength ratings, consult tables for the WLL of chain to be used. When accuracy is not critical, the use of the following rule of thumb will be an indicator of the WLL.

Rule of Thumb for Chains	Example
1. Convert the chain diameter to eighths. 2. Square the numerator. 3. Multiply by 600 for alloy steel chain. 4. = approximate WLL in pounds.	To find the WLL of a 3/8-inch alloy steel chain, perform the following calculation: 1. The 3/8-inch diameter already has a denominator expressed in eighths. 2. Square the numerator 3^2: = 9. 3. Multiply 9 by the rule-of-thumb factor of 600. 4. The WLL is, therefore, $600 \times 9 = 5{,}400$ pounds.

Specific Hazards when Working with Chain

Hazards	Barriers
Serious injuries have occurred when a chain has broken and snapped back, especially when using it for unknown weight (tension), like pulling a truck out of mud.	A common perception in the field is that a chain is the strongest piece of rigging for towing or lifting something very heavy. The safety rule of staying out of the direct line of a pull does not work well when the windshield and the driver of the stuck vehicle are in direct line. A chain should be used only if the load is known. Avoid using a chain for towing a stuck vehicle. In case of a chain failure, the chain ends can recoil with lethal force.
A chain sling to be used for hoisting may have been weakened when used for other duty.	Chain slings used for hoisting must not be used for any other purpose. For example, a chain used to bind a load can be subject to severe shock loading.
Sharp bends can weaken or damage chains.	Padding must be used when using chain slings on steel towers. Never use a knot in a chain. Use the proper end fittings.

15.2.4   Ratchet Chain Hoist

Like all hoisting devices, chain hoists require scheduled maintenance and testing by a competent mechanic. The lower hook is the weakest part and will start to spread under overload. A chain hoist with a bent hook or binding clutch should be removed from the field and repaired.

Do not do any of the following:

- Use a chain hoist to bind poles on a trailer.
- Tie the chain into a half hitch behind the head of an anchor rod.
- Overload a chain hoist by using a cheater to lengthen the handle.
- Leave a chain hoist under tension for prolonged periods.

15.2.5 Web Hoist

Web hoists were originally intended and reserved for work on or near live circuits. If a web hoist is being used for general duty such as pulling a guy, it should be identified so that it will not be used on a live circuit. The nylon material is vulnerable to contamination and must not be connected directly between a live conductor and a structure without an insulated link.

15.2.6 Snatch Blocks

Bisect Tension Angle

It is necessary to know the bisect tension that a snatch block will be holding, the WLL of the block, and the WLL of the anchor point. When calculating or using a rule of thumb to determine a bisect tension on a block, ensure that the correct angle is measured. Figure 15–2 shows the two typical ways an angle can be measured.

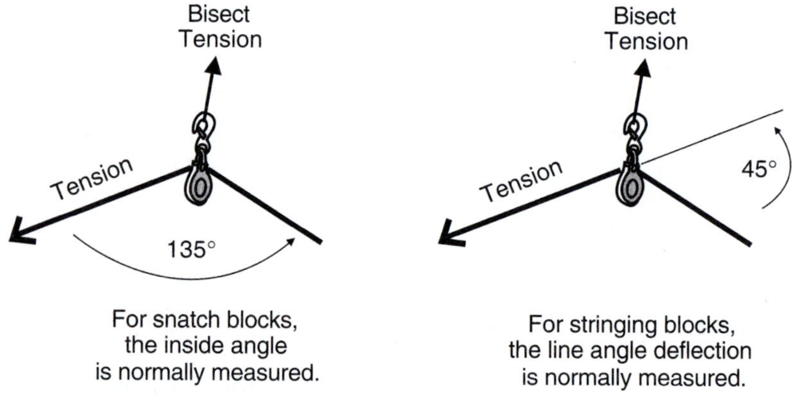

Figure 15–2   Two ways to measure an angle at a block.

A Rule of Thumb for Bisect Tension on a Snatch Block

The rule of thumb for determining the bisect tension on a snatch block and anchor point by a wire rope is to use the deflection angle illustrated in Figure 15–3.

The strain on a block and anchor point

- 0.76 × rope tension for 135°
- 1.0 × rope tension for 120°
- 1.41 × rope tension for 90°
- 1.73 × rope tension for 60°
- 1.84 × rope tension for 45°
- 2.0 × rope tension for 0°

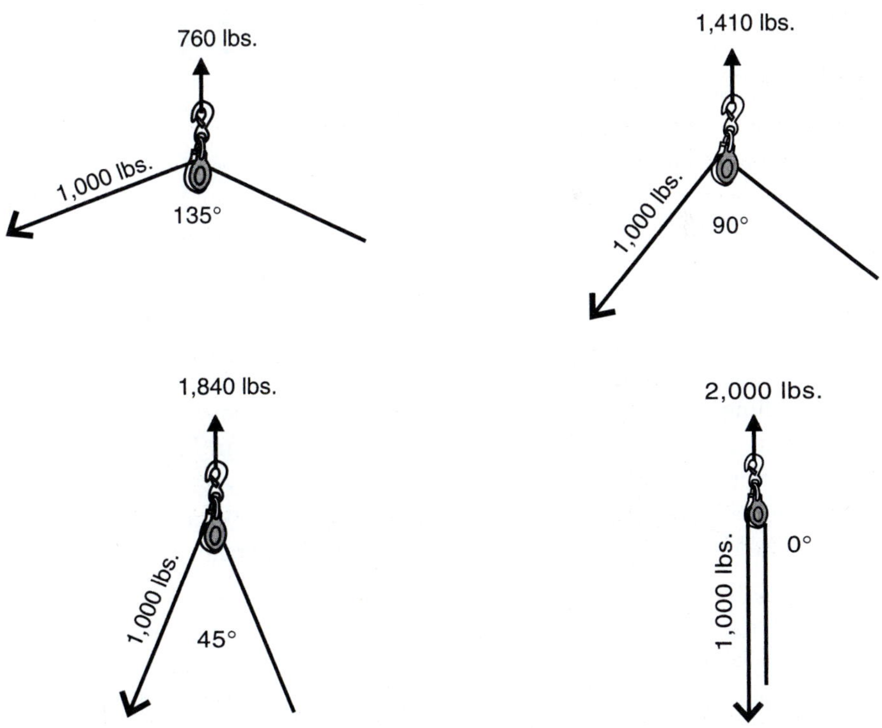

Figure 15–3   Load on a snatch block.

A Rule of Thumb for Matching Rope Blocks and Rope Size

An improperly sized block can derate the WLL of a wire or fiber rope. The following provides a rule of thumb for determining the proper size of a block:

- The snatch block should have 1 inch of shell diameter for every 1/8 inch of fiber-rope diameter.

- The snatch block should have 1 inch of shell diameter for every 1/16 inch of wire-rope diameter.

Friction Using Blocks

There is a loss due to friction any time a rope, wire, chain, and such goes through a block. When more than one block is used, such as the sheaves in a set of tackle blocks or the stringing blocks over many spans, the friction loss can become significant. Figure 15–4 shows some typical values that can be used when calculating friction.

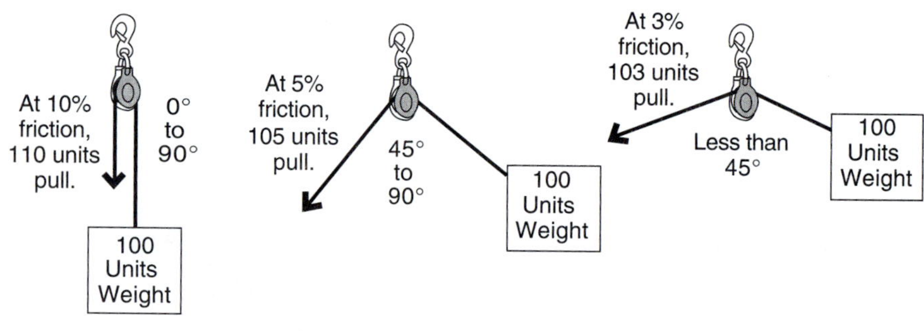

Figure 15–4   Friction on a block.

15.2.7 Anchor Pulling Eye

When pulling a down guy, the limiting factor could be the anchor-pulling eye being used. The two types of anchor-pulling eyes shown in Figure 15–5 each have safe working loads of 3,000 pounds (1,400 kg). It is very easy to overload the pulling eye, grip, or chain hoist when pulling on a down guy that is in service. Use a derrick boom to help hold or adjust the pole rake.

15.2.8 Collapsible Bull Wheel

The WLL for a typical collapsible bull wheel (butterfly) (as shown in Figure 15–6) is 4,000 pounds (1,800 kg). When used on the extension shaft of a typical digger derrick boom-tip winch, the WLL is reduced to 800 pounds (360 kg). When taking up a rope under tension, the compressive force of a stretched rope can collapse the bull wheel. The number of turns around the bull wheel should be limited to four. Workers should keep a safe distance and not stand in direct line with a bull wheel.

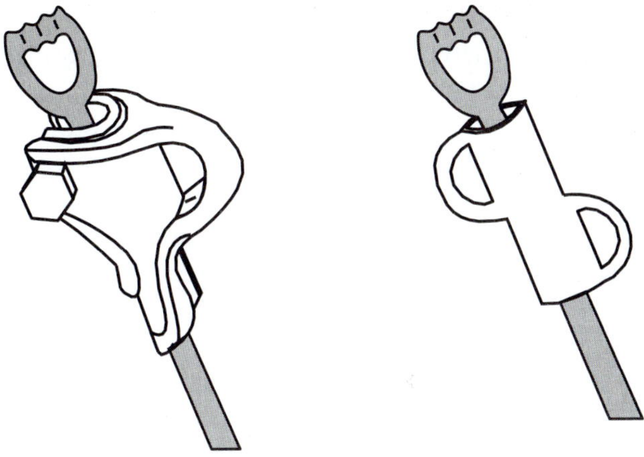

Figure 15–5 Anchor pulling eyes.

Figure 15–6 A collapsible bull wheel.

15.2.9 Gins

Many types of gins and cargo booms are used in line work, including transformer gins, arm gins, and pole gins. Gins are designed for vertical loads only, so any side pulls and tagging out of a load will derate the WLL of the gin. The load line should be parallel to the gin pole.

Figure 15–7 shows how the load on a gin is doubled. The transformer gin has a 2,000-lb. maximum rating, including load and fall-line pull. When calculating the load that can be lifted (allowing 10 percent for friction) on a single line, the maximum load is 900 pounds.

15.2.10 Hand Line

Lessons learned.

- Keep the hand line away from traffic.

- Do not tie the hand line to a truck.

- While climbing, attach a hand line to a worker's belt so that it breaks away if it gets snagged by passing vehicles or other equipment. Belt

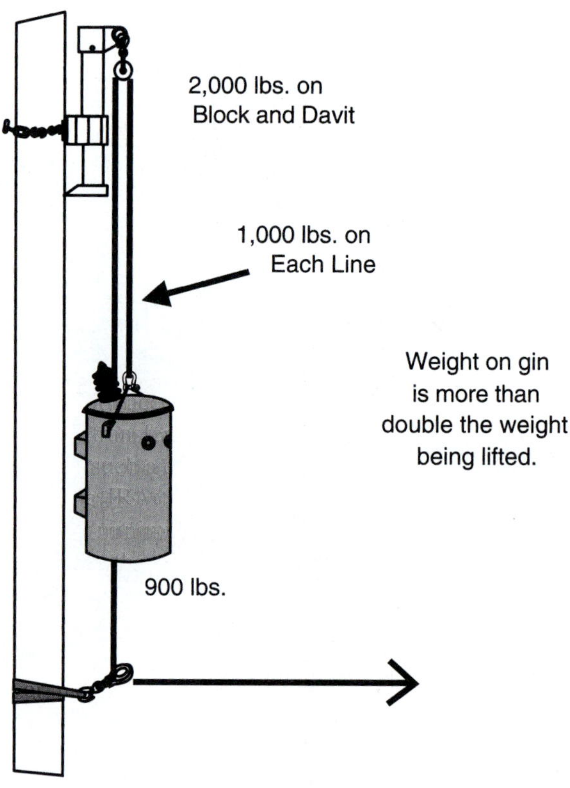

Figure 15–7   A transformer gin.

hooks that open and release their load when under too much weight are available.

- A hand line is a hoisting device that is also used by many employers as a pole-top/tower-top rescue device. For use as a rescue device, it should be made with hardware and rope capable of withstanding a person's weight plus a 10-to-1 safety factor.

- Figure 15–8 shows how the weight on the anchor point is double the weight being hoisted.

15.2.11 Capstan Hoist Hazards and Facts

- To control the weight to be lifted by a capstan hoist (Figure 15–9), the number of wraps around the capstan hoist should be kept so that about 20 to 40 pounds (9 to 18 kg) of pull are on the fall line. More turns will mean a loss of control to stop the load.

- Trying to add or remove turns during a lift could cause a loss of control.

- To determine the number of wraps needed for the load to be lifted, use data specific to the hoist being used, as shown in Table 15–1.

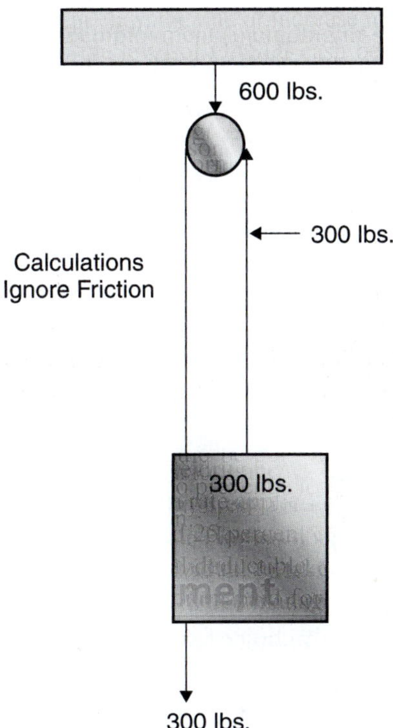

Figure 15–8 A vector representation of a hand line.

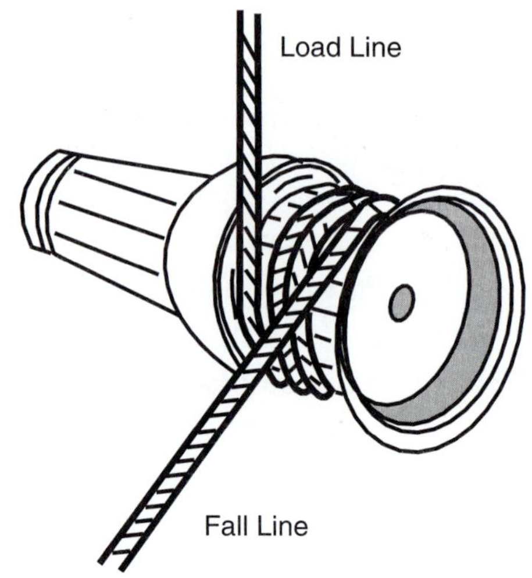

Figure 15–9   A capstan hoist.

TABLE 15–1   Rope Turns on the Drum for Weight to Be Lifted

Rope Turns on Drum	Ratio of Load Line to Fall Line	Weight on Load Line to Be Lifted Pounds (kg) Pull on Fall Line		
		20 lbs. (9 kg)	*30 lbs. (14 kg)*	*40 lbs. (18 kg)*
3	1:20	400 (180)	600 (270)	800 (360)
3.5	1:30	600 (270)	900 (400)	1,200 (540)
4	1:50	1,000 (450)	1,500 (680)	2,000 (900)
4.5	1:70	1,400 (630)	2,000 (900)	2,800 (1,270)
5	1:100	2,000 (900)	3,000 (1,360)	4,000 (1,800)

- Letting the capstan hoist turn without advancing the rope can cause the rope to melt or weld to the drum, where it will start to wrap up the rope like a winch.

- Any hoist used to raise workers up to a conductor for specialized barehand work must be kept exclusively for that work, and the worker should use a backup fall arrest system.

15.2.12 Rope Blocks

Rope blocks are often the only choice for work in back lots, islands, and at wide ditches.

Figure 15–10 shows how a 1,000-pound weight is pulled up with only 500 pounds on the fall line. It also shows that the weight on an anchor point is 1,500 pounds instead of the 2,000 pounds it would have been without the ex-

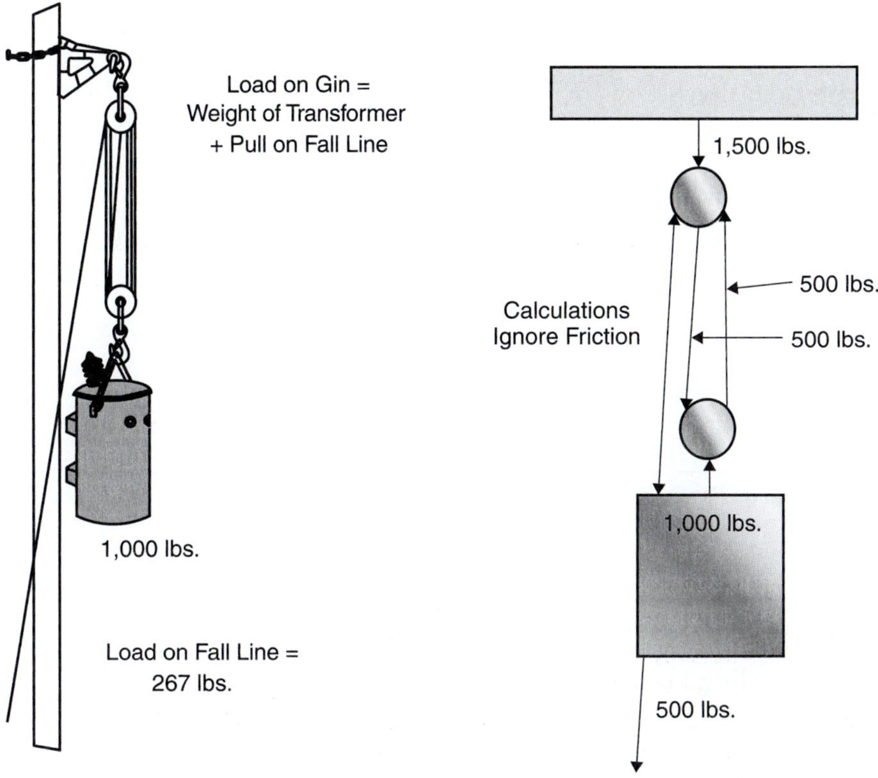

Figure 15–10   Mechanical advantage of rope blocks.

tra line and block. More lines and block sharing the weight will reduce the pull on the fall line and therefore the weight on the anchor.

The Mechanical Advantage of Rope Blocks

The mechanical advantage of rope blocks is shown in the following formula:

$$Fall\ Line\ Load = \frac{Load + (Numbers\ of\ Sheaves \times 10\%\ of\ Load\ for\ Friction\ Loss)}{Number\ of\ Lines\ from\ Moving\ Block}$$

For example, the 1,000-pound transformer being raised with a set of three sheave blocks in Figure 15–10 has six lines moving from the moving (bottom) block.

$$Load\ on\ the\ Fall\ Line = \frac{1,000 + (6 \times 0.1 \times 1,000)}{6} = 267\ pounds$$

The friction through the blocks will vary with maintenance and type of bearings.

Matching Rope Blocks and Rope Size

The shell of the block should be eight times the diameter of the rope. For example, a 1/2-inch (1.3-cm) rope requires 8 × 0.5-inch = 4-inch (10-cm) blocks.

WLL of Rope Blocks

The WLL of a set of blocks will depend on the WLL of the blocks and the size and type of rope. Table 15–2 shows the WLL of rope blocks for polypropylene rope.

TABLE 15–2   WLL of Rope Blocks

Typical WLL of Blocks at a 9-to-1 Safety Factor		
Polypropylene Rope	*Set of 2-Sheave Blocks*	*Set of 3-Sheave Blocks*
1/2 in. (1.3 cm)	1,500 lbs. (700 kg)	2,000 lbs. (900 kg)
3/4 in. (1.9 cm)	2,400 lbs. (1,000 kg)	3,000 lbs. (1,400 kg)

15.2.13 Sling Hazards and Facts

- The strength of a sling depends on its material strength and the manner in which it is hitched to the load (Figure 15–11).

- A single-leg vertical sling with proper end fittings is rated at the strength of the rope and the type of end fittings.

- A tag line is needed when lifting with a single vertical sling because of the tendency for the load to rotate and swing. Allowing the load to rotate can untwist a sling and weaken any hand-tucked eyes.

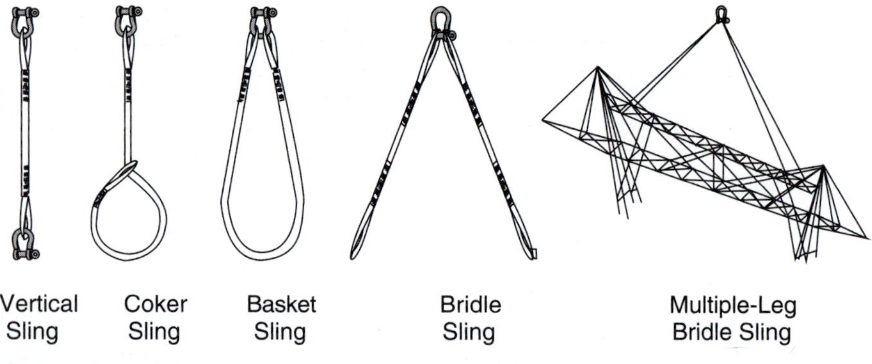

Figure 15–11   Connecting to a load with a sling.

- A single-wrap choker hitch is the weakest form of a hitch. When it is not under a load, the hitch can open and release its load. Forcing the choker end down places a sharp bend in the sling and reduces the sling angle. Figure 15–12 shows pitfalls and some of the complexities when using a choker.

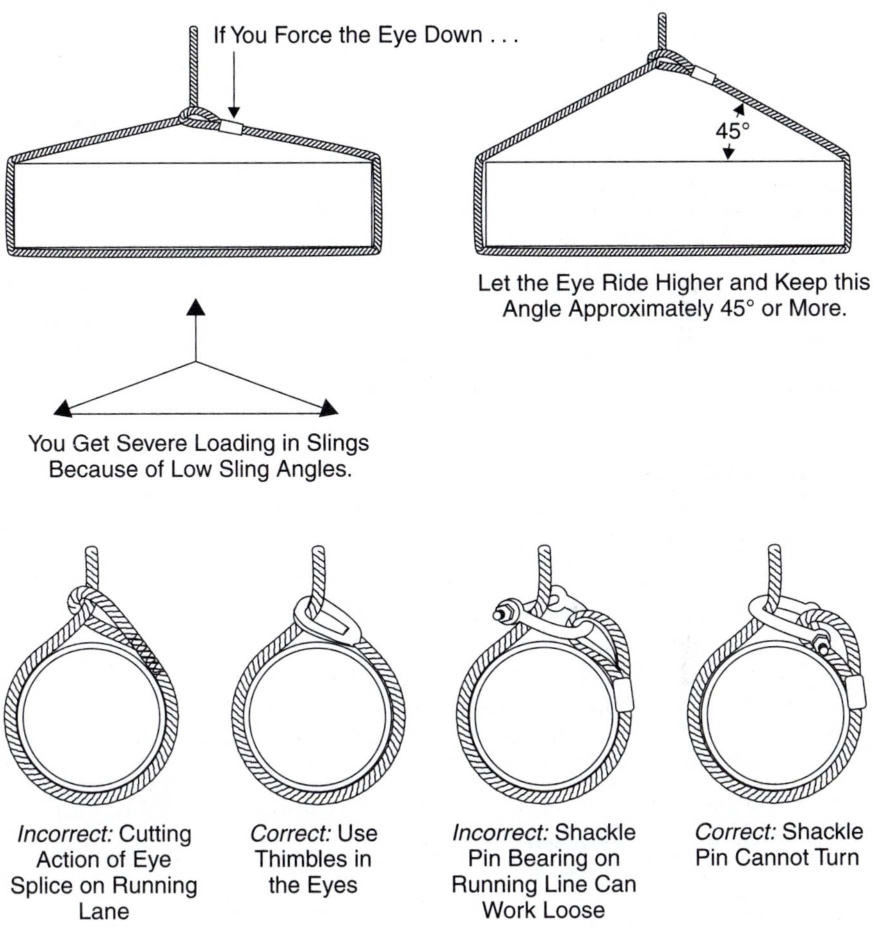

Figure 15–12   Pitfalls of using a choker hitch.

- A *basket hitch* should only be used on straight lifts. Moving the load within the basket hitch can damage and weaken the sling.

- A *bridle sling* is intended to be a two-leg sling with two legs carrying the load. If a single sling is used through a clevis and the two ends are attached to a load, an unbalanced load can cause the heavy end to drop and the sling to slide through the clevis.

- When lifting objects with *multiple-leg slings* (three or four legs), two of the legs should be capable of supporting the total load. Multiple-leg slings must be selected to suit the most heavily loaded leg instead of the total weight.

- A key factor in determining sling stress is the resultant sling angle when making a lift. Note how the sling angles in Figure 15–13 affect the stress on the sling. For most distribution-line work, if a sling is chosen with a WLL of double the weight to be lifted, it will not be unwieldy and the sling angle can be ignored.

- Do not form the end of a boom-tip winch into a sling to make a lift. The sling angles produced will overstress the rope winch and definitely shorten its life.

- Sudden starts or stops place much heavier loads on rigging. A load could be increased by two to fifty times its actual weight. When lifting a load, the lift should be started very slowly until the sling becomes taut. Then lifting should continue slowly until the load is suspended. The load must be prevented from rotating. Use a tag line to prevent any rotation and to reduce swinging.

Using Web Slings

Web slings are preferred for distribution work such as lifting transformers, regulators, and such because they are more resistant to cutting and abrasion than fiber rope and easier to work with than wire rope or chains (see Figure 15–14).

The sling size and rated load are identified on a label, which is usually a sewn-on-leather tag. The WLL is based on the configuration that is used. Slings should be discarded if any of the following is true:

1. The label showing the load rating is not visible or missing.

2. There are holes, tears, cuts, or snags; the wear indicator (red strand) is visible; or there is excessive abrasion.

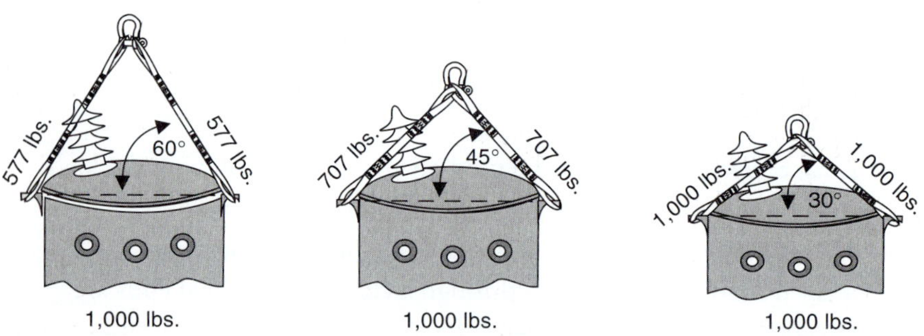

Figure 15–13   Sling angles.

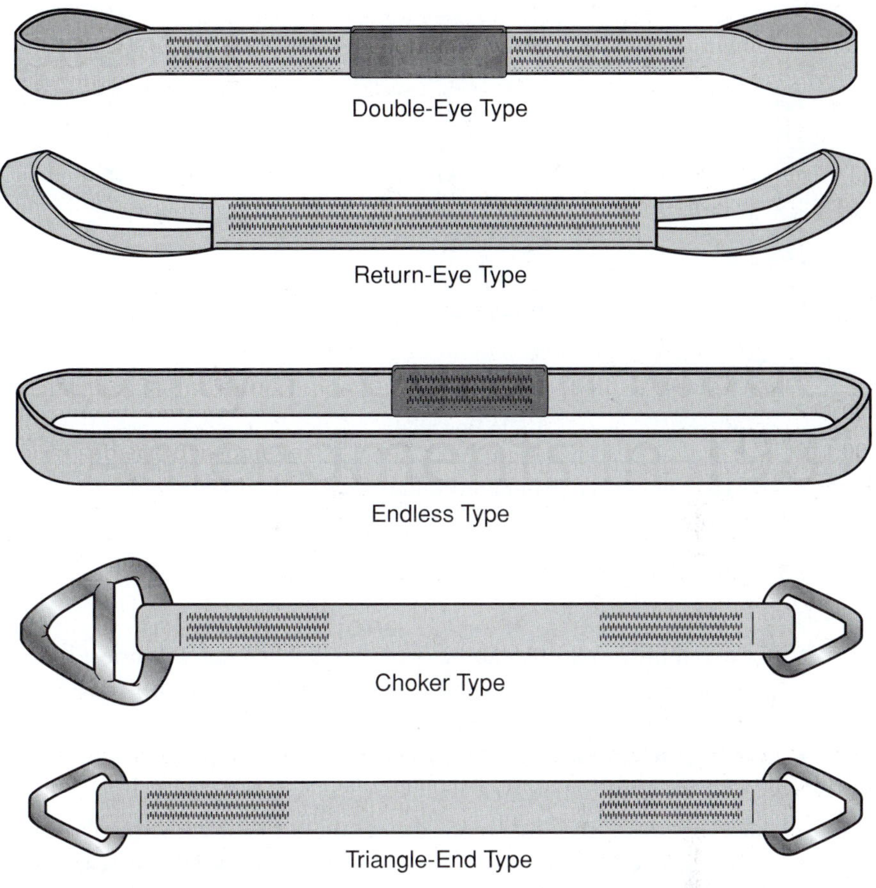

Figure 15–14 Web slings.

3. Knots are present in any part of the sling.

4. Excessive color has been lost, which indicates ultraviolet light damage.

15.3 Working with Tensioned Conductors

15.3.1 Calculating Conductor Weight in a Span

1. On level terrain, the weight of a conductor =

   Conductor Weight in lbs./ft. (kg/m) $\times$ *(1/2 Span A + 1/2 Span B)*

2. When the bottom of the sag is not midspan, as seen in Figure 15–15, the weight of a conductor =,

   Conductor Weight in lbs./ft. (kg/m) $\times$
   (Length to Lowest Sag Point in Span A + Length to Lowest Sag Point in Span B)

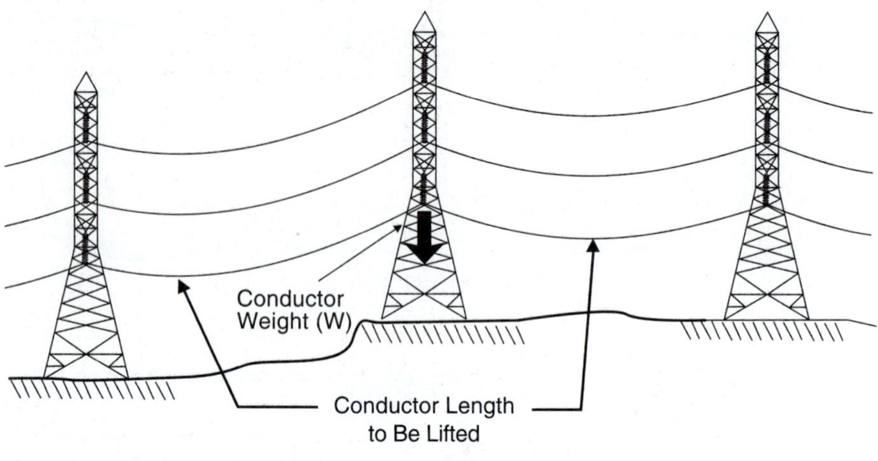

Figure 15–15   The effect of a hill on conductor weight on a structure.

- An aerial lift or rigging can be overloaded when consideration is not given to lifting a conductor from a structure on a hill or lifting a conductor to a higher position.
- A conductor's weight doubles when it is lifted to its maximum height, which occurs when the lowest point of sag reaches an adjacent structure. Therefore, as a conductor is lifted with an aerial device or live-line tools to a higher location, the weight increases as the lowest sag point moves farther away.

15.3.2 Calculating Conductor Tension

Conductor Tension =

$$\frac{\textit{Conductor Weight in lbs./ft. (kg/m)} \times \textit{Span Length}^2 \textit{ in ft. (m)}}{8 \times \textit{Sag in ft. (m)}}$$

Some Rules of Thumb for Conductor Tension

1. The tension on a conductor is generally the heaviest weight dealt with by powerline workers.

2. On distribution lines:

   - Line tension will be less than 500 pounds (225 kg) for 3/0 or smaller and less than 750 pounds (340 kg) for 3/0 and larger aluminum conductor.

- Line tension will be less than 750 pounds (340 kg) for 3/0 or smaller and less than 1,000 pounds (450 kg) for 3/0 and larger copper conductor.

3. At temperatures below freezing, the conductor tension increases 20 percent for each 20°F (10°C) drop in temperature.

4. At temperatures above freezing, the conductor tension increases 10 percent for each 20°F (10°C) drop in temperature.

5. When pulling up on a conductor, the tension increases as the sag decreases. Line tension is doubled when one-half of the sag is removed. Line tension is tripled when two-thirds of the sag is removed.

15.3.3 Calculating Tension on a Loaded Down Guy

$$Guy\ Tension = T \times \frac{\sqrt{L^2 + H^2}}{L}$$

where   H = height of the guy attachment
   L = length of the guy lead
   T = tension of the conductors held by the down guy

15.3.4 Calculating guy tension of the example in Figure 15–16.

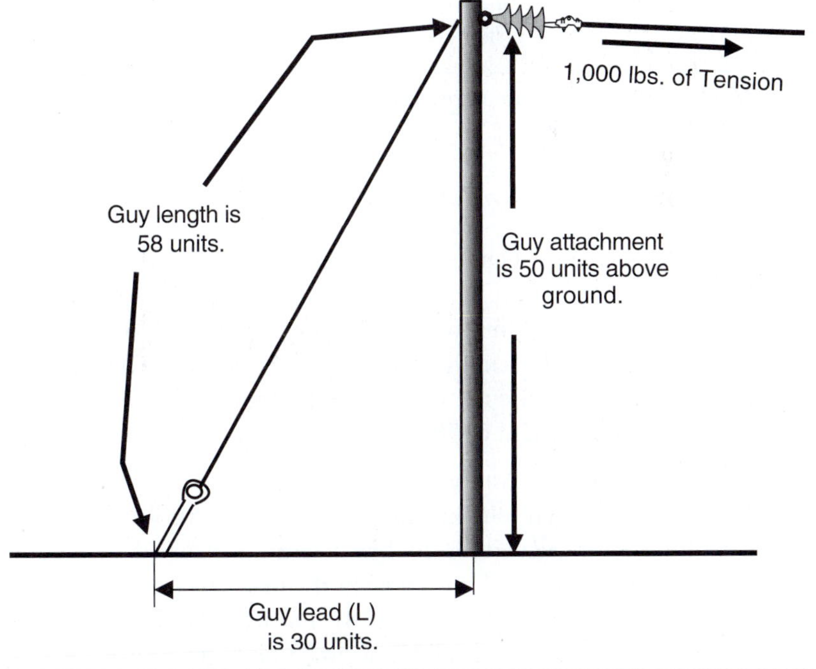

Figure 15–16   Calculating guy tension.

$$Guy\ Tension = 1,000 \times \frac{\sqrt{30^2 + 50^2}}{L} = 1,944\ lbs.$$

15.3.5 Calculating Vertical Load Applied by a Down-Pull

The vertical load applied to a crossarm, snatch block, or a stringing block by a down-pull (Figure 15–17) can be excessive when stringing conductor or temporarily dead-ending conductor to the ground. A short lead will exert a high vertical weight, which could overload a structure or its components.

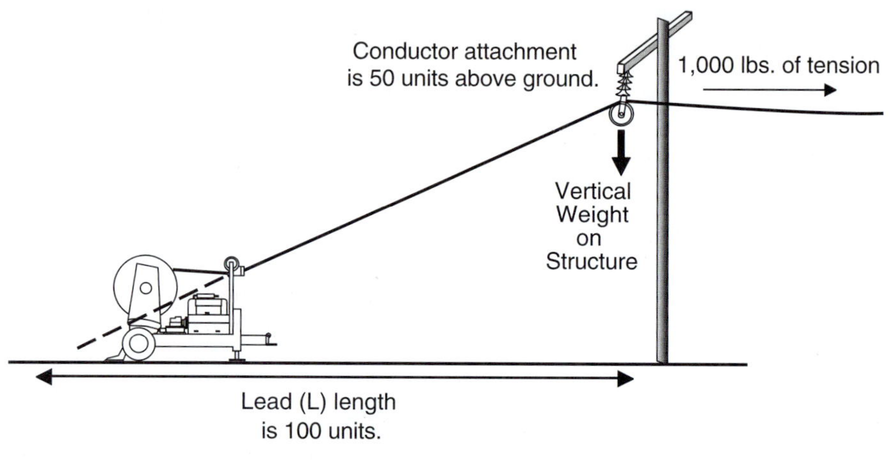

Figure 15–17   Vertical load on a structure.

$$Vertical\ Weight\ on\ a\ Structure = \frac{T \times H}{L}$$

In the example shown in Figure 15–17, T = 1,000 pounds, H = 50 feet, and L = 100 feet.

Therefore, the vertical weight on the crossarm = (1,000 × 50)/100 = 500 pounds. It is best to keep the length (L) as long as practical during stringing in the preceding example because the vertical weight would be only 100 pounds if the length (L) were increased to 500 feet.

15.3.6 Measuring a Line Angle in the Field

A line angle must be known to conform the proper framing for the structure, as well as be capable of calculating the bisect tension involved in handling a conductor. To measure an actual line angle in the field, measure out 57 feet in the two directions (as shown in Figure 15–18). The length of Line X represents the line angle. This method is reasonably accurate up to 45 degrees.

For example, if Line X is 30 feet, the line angle is 30 degrees. (For metrics, measure 6 meters in each direction. The line angle is the length of Line X in meters × 10. For example, if Line X is 3 meters, the line angle is 10 × 3 = 30 degrees.)

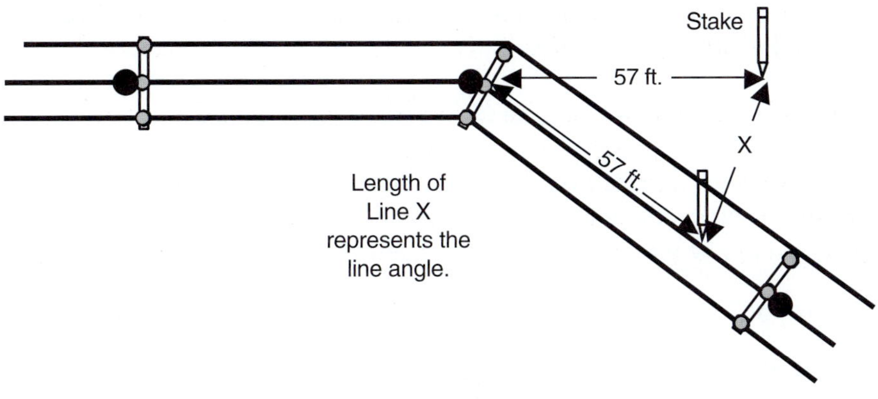

Figure 15–18 Measuring a line angle.

15.3.7 Calculating the Bisect Tension of a Conductor at a Corner

Bisect tensions should be known when a conductor is lifted or relocated on a corner structure. The approximate bisect tensions of a conductor can be calculated using Table 15–3.

TABLE 15–3 Factors to Calculate Bisect Tension

The table shows how the factors are used to calculate the bisect tensions various angles given 1,000-pounds line tension. Using the table, an actual line tension can be inserted in the proper row to calculate a bisect tension.

Line Angle (degrees)	Bisect Tension (lbs. or kg)	= Line Tension	× Factor
90	1,500	1,000	1.5
75	1,250	1,000	1.25
60	1,000	1,000	1.0
45	750	1,000	0.75
30	500	1,000	0.5
15	250	1,000	0.25

If a more accurate calculation of the bisect tension is needed, the bisect tension can be calculated as follows:

$$Bisect\ Tension = \frac{Line\ Angle}{60} \times Line\ Tension$$

Note: The bisect tensions on the snatch blocks in section 15.2.6 may appear different; however, the angle of the rope is measured inside the block, while line angles tend to be measured as shown in Figure 15–2.

Moving a Conductor into a Corner

The sag and tension of a conductor can change very quickly when moving a conductor into a corner. The tension and sag increase can cause a midspan contact with live overbuilt conductors, a broken jib, or bucket truck instability. Calculating the bisect tension ahead of time is complex. Keeping a close watch on the sag as the conductor is moved in and out of a corner is the most practical way to prevent excess tensions.

Controlling Tree Work Hazards

16.1 Tree Work Hazards Exposure

Tree Work Hazards	Controls
1. Arborists have a high exposure to musculoskeletal injuries because of climbing, hauling brush, and picking up branches.	Pruning and removing trees are very physical tasks. In addition to keeping fit, every arborist should participate in a back-care program.

(continued)

16.1 Tree Work Hazards Exposure *(continued)*

Tree Work Hazards	Controls
2. Arborists have a high exposure to repetitive strain injuries, especially from using pruners and pole saws while leaning out of buckets.	In addition to tool and bucket design, an arborist must be part of an education and exercise program that reduces repetitive strain injury.
3. Arborists have a high exposure to severe cuts when using chain saws, probably the highest-risk power tool ever invented.	Arborists use more personal protective equipment than any other electrical utility worker. In addition to a hard hat and protective footwear, when operating a chain saw an arborist also wears leg protection, eye protection, hearing protection, and protective gloves.
4. When climbing trees or working from aerial devices, arborists have a high exposure to falls from heights.	Specialized skills, equipment, and training are necessary for climbing trees safely and minimizing the risks of falling. The risk of tree climbing can be reduced by using a fall protection system. The one shown in Figure 16–2 later in this chapter is a belaying technique requiring very little in equipment. This method can be used when climbing and must be done to secure a rope for tree removal.
5. Arborists have a high exposure to electrical hazards. Electrical contact causes about 30 percent of all fatalities related to tree work. Direct contact with a powerline or a tree branch touching a powerline increases the risk of receiving a fatal electric burn.	Arborists must be able to recognize the components of the electrical system being worked on and must have extensive training on the proper techniques for pruning and removing trees near live circuits.

16.1.2 Danger Tree Definitions

The U.S. Federal Occupational Safety and Health Act (OSHA) regulations have a rule regarding danger trees: "Each danger tree shall be felled, removed, or avoided. Each danger tree, including lodged trees and snags, shall be felled or removed using mechanical or other techniques that minimize employee exposure before work is commenced in the area of the danger tree." A danger tree includes any standing tree that presents a hazard to employees due to conditions such as, but not limited to, deterioration or damage to the tree, and direction or lean of the tree.

Signs of a danger tree can be:

1. Large, dead branches and hangers in the tree and on the ground

2. Cracks or splits in the trunk or where branches are attached

3. Cavities or rotten wood along the trunk or in major branches

Other regulations require a utility to remove any diseased or unstable trees that are tall enough to fall into a powerline. In a utility, this type of tree may also be called a *danger tree.*

16.2 Tree Work Near Electrical Circuits

16.2.1 Minimum Approach Distances for Tree Pruning

Occupational safety regulations do not allow an unqualified person to come within 10 feet (3 m) of a distribution electrical circuit and an additional 4 inches (10 cm) for every 10 kilovolts over that, which works out to 14 feet (4.3 m) for 169 kilovolts, 16 feet (4.9 m) for 230 kilovolts, and 25 feet (7.6 m) for 500 kilovolts. Working near electrical circuits is the highest-risk hazard that an arborist must control. The most common and most effective barrier against electrical contact is maintenance of a space between the work zone and the electrical conductors. The OSHA regulations specify the minimum approach distances. These are *minimum* distances for qualified arborists—not necessarily safe distances. While the length of an insulated and regularly tested pruner is electrically safe for the distance specified, an inadvertent movement or a limb bridging across the pruner can cause electrical contact. Powerline workers should plan to stay farther away than the minimum distance recommended, possibly another 3 feet (1 m). When it is necessary to work *at or close to* the minimum distance, treat that like an exception requiring additional planning, such as the following:

1. Stop and plan your approach and movement, preferably keeping the hazard at arm's length.

2. Plan the work so that your exposure time will be minimized.

3. Discuss your plan with a person who will act as a dedicated observer for the duration of the work.

TABLE 16–1 Minimum Working Distances for Tree Pruning

Voltage (Phase to Phase)	Minimum Working Distance
2.1–15 kV	2 ft., 0 in. (0.6 m)
15.1–35 kV	2 ft., 4 in. (0.7 m)
35.1–46 kV	2 ft., 6 in. (7.6 m)
46.1–72.5 kV	3 ft., 0 in. (0.9 m)
72.6–121 kV	3 ft., 4 in. (1 m)
138–145 kV	3 ft., 6 in. (1.07 m)
161–169 kV	3 ft., 8 in. (1.1 m)
230–242 kV	5 ft., 0 in. (1.5 m)
345–362 kV	7 ft., 0 in. (2.1 m)
500–552 kV	11 ft., 0 in. (3.4 m)
700–765 kV	15 ft., 0 in. (4.6 m)

16.2.2 Tree Work Electrical Hazards and Controls

Hazard	Controls
Encroaching on the minimum approach distance to a live conductor.	The minimum approach distance applies to a live conductor, switch, etc., as well as to a branch or limb in contact with a live conductor. If the work cannot be done while maintaining a minimum approach distance, then obtain an outage or use cover-up.
Encroaching on the minimum approach distance working with a pruner or pole saw.	While the length of an insulated and regularly tested pruner is electrically safe for the distance specified, an inadvertent movement or a limb bridging across the pruner can cause electrical contact. Keep the bucket away from second points of contact and routinely use more pruner length than the minimum approach distance.
Climbing a tree in contact with a live conductor exposes a climber to tree shock.	Before going aloft, ensure that no limbs are in contact with the tree. If a limb is making contact, do not climb the tree. A limb or branch can be removed or cut using an insulated pruner, working from an insulated aerial device, and while not making bodily contact with any part of the tree. A tree is not an excellent conductor, and no voltage drop would occur as the current flows through the length of the tree. The voltage is

(continued)

16.2.2 Tree Work Electrical Hazards and Controls *(continued)*

Hazard	Controls
	highest at the contact or entry point and lowest at the exit point, usually at ground level. As shown in Figure 16–1, the hands on the tree could be at a different potential from the feet because the voltage drops as it flows through the resistance of the tree.

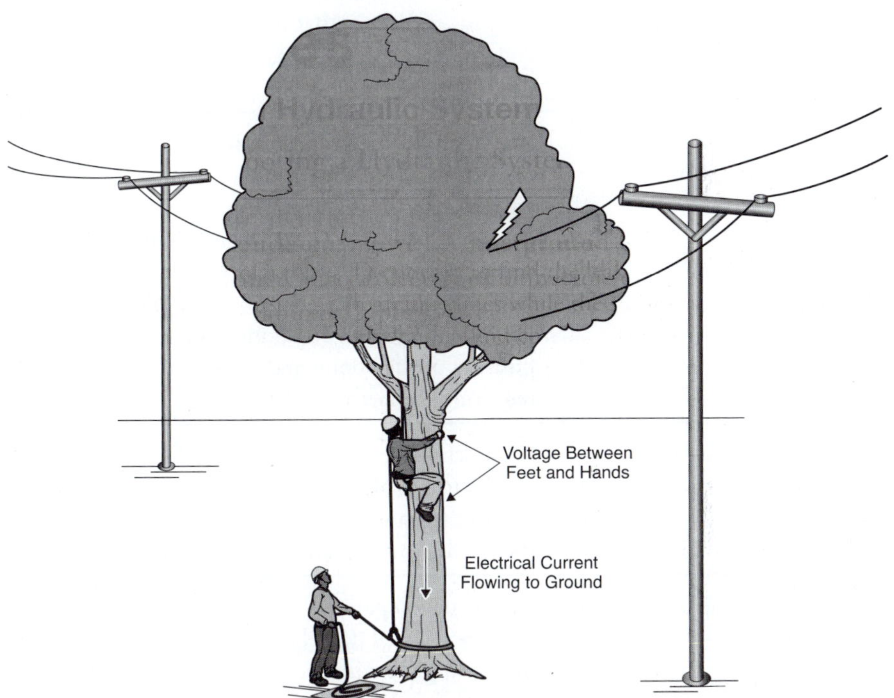

Figure 16–1   Tree shock.

| A climber putting weight on a limb and causing the limb to make electrical contact. | Climb on the side of the tree farthest from the conductor and face toward the hazard to keep it visible. Choose a crotch to tie into that would tend to swing you away from the conductor in case of a slip or fall. |

(continued)

16.2.2 Tree Work Electrical Hazards and Controls *(continued)*

Hazard	Controls
Removing a branch in touching a live conductor.	When working from a bucket, it is standard procedure to use an insulated pole pruner to remove a branch lying on a live conductor (hanger). This can be done on voltages up to 15 kV. On higher voltages, a limb contacting two conductors might short out the circuit. A short circuit, especially near a substation, will be very explosive. Use eye protection and the full length of the pruner. Ensure that no one stands under the conductor in case it burns off and falls to the ground.
Limbs and branches being dropped short-circuits the lower boom insert.	While the probability is low for the lower boom insert to be short circuited by a branch coincident with the lower boom making contact with a live circuit, the consequences for anyone touching the truck are high. If it is necessary to contact the truck, the person on the ground should ask the operator to stop all boom movement, check the position of the boom, and then do the task requiring contact with the truck.
Working above live brittle conductors.	A branch can easily break a #4 ACSR or #6 copper conductor. While it is always a good practice to stay out from under the circuit, it is critical when the conductor is small and has low tensile strength.
Electrical tracking on a pruner or pole saw.	The insulated parts of the pruner and pole saw must be cleaned daily before use. Wiping down the insulated parts with methyl hydrate will ensure that the tools are dry. Unless the rope in the pole pruner is live-line rope and tested and treated as such, the insulated link in the rope should be cleaned daily. The pruner, pole saw, and hydraulic hose (if equipped) should be scheduled regularly for an electric test and also for replacement of any parts that affect the insulated parts.
Felling a tree within falling distance of a line.	Rope the tree.

16.3 Operating a Chain Saw

16.3.1 Personal Protective Equipment when Using a Chain Saw

In addition to a class E hard hat and safety footwear, a chain-saw operator wears the following:

- *Safety glasses and/or face shields* are worn to protect the face from objects propelled toward the face by the saw.

- *Hearing protection* is required because a chain saw with a noise level of less than 85 decibels has not been invented. Many types of hearing protection are available, including foam plugs and ear muffs. Hearing protection with the highest Noise Reduction Rating (NRR) provides better protection.

- *Chaps or cut-resistant material,* such as ballistic nylon sewn into special chain-saw pants, will reduce the risk of cuts to the legs. *Leg protection* covers the full length of the thigh to the top of the boot on each leg.

- *Chain-saw gloves* are specially constructed to protect the back of an operator's hand should the saw kick back.

16.3.2 Chain-Saw Operating Rules

The following regulations regarding chain-saw use reflect the high risk of contacting a moving chain-saw blade:

1. The power saw operator wears head, eye, hearing, hand, foot, and leg protection.

2. The saw is equipped with a chain brake and a protective device that minimizes chain-saw kickback. Kickback happens when the upper portion of the saw tip comes in contact with another object, causing the saw to jump or kick back toward the operator.

3. The saw is well-maintained and the chain is sharp. The clutch is adjusted so that the clutch will not engage the chain drive at idling speed and is equipped with a continuous-pressure throttle-control system that will stop the chain when pressure on the throttle is released.

4. The saw engine is stopped for all cleaning, refueling, adjustments, and repairs, except for adjustments that cannot be done without the engine running.

5. The saw is started while it is held down firmly on a solid surface, such as the ground or a stump, and the chain brake is engaged.

6. Drop starting of a saw can be done safely only from outside the bucket of an aerial device.

7. When starting or operating a saw, no one is within 6 feet (2 m) of the saw operator. The saw is started at least 10 feet (3 m) away from the fueling area.

8. The saw is not operated with only one hand.

9. Loose material that may catch the saw is removed.

10. A saw weighing more than 20 pounds (9 kg) is never used above the height of an operator's shoulder, in an aerial basket, or in a tree. One reason why safety rules prevent a saw from being used above the shoulders is because a kickback is very hazardous in this position.

11. In an aerial bucket, the saw is stored in a sheath when it is not in use.

12. When working in a tree, except when working from an aerial bucket, a power saw weighing more than 15 pounds (6.8 kg) should be supported by a separate work rope tied in such a way that it will allow the saw to swing clear of the operator.

13. A power saw should not be raised or lowered from a tree with the motor running.

14. In addition to a chain brake, the risk of kickback is reduced when a sharp, properly tensioned "low kickback chain" is used and the saw is held firmly with two hands—and not while standing over the saw.

16.3.3 Freeing a Trapped Saw

If a saw becomes trapped while you are pruning with it, switch off the saw and, if not already attached, attach it to a separate tool line. Try lifting the trapped branch and, with the chainsaw brake off, pull the saw from the kerf. Alternatively, use a hand saw to release the trapped saw by making a release cut outward toward the tips of the limb, close enough to allow the undercut made previously with the chain saw to be effective.

16.4 Tree Climbing and Tree Rescue

16.4.1 Tree Climbing Procedure

Step	Action	Details
1	Prepare and discuss a job safety plan.	Typical job safety planning includes identifying such high-risk hazards as electrical contact, danger tree hazards, and falling hazards, then planning to control for these hazards.
2	Inspect the tree.	1. Identify the location and voltage of any nearby powerlines and places for any limb contact with the tree.

(continued)

16.4.1 Tree Climbing Procedure *(continued)*

Step	Action	Details
		2. Choose the location for final crotching in.
		3. Inspect the tree for dead and broken limbs, splits, and decay to ensure that the tree is not classified as a danger tree.
		4. Check for wasp, bee, and hornet nests.
3	Choose the first and final crotch.	The crotch is the anchor point for safe climbing and for anchoring an arborist in various working positions while descending. Ideally the crotch would be directly above the work area, in a position where a slip would swing the arborist away from any powerlines. The rope is passed over a branch and around the trunk as high above the ground as possible using branches with a wide crotch. Tight V-shaped crotches should be avoided because they can bind the rope as well as feet and hands.
		A false crotch can be used in place of a natural crotch. A false crotch (friction saver or cambium saver) is a nylon strap with metal rings at each end, one larger than the other. The climbing rope is threaded through the two rings. A false crotch allows a rope to be placed anywhere on a trunk or branch.
4	Crotch the rope.	Climbing a tree safely requires getting the climbing rope into the tree over a suitable crotch that will serve as an anchor. A lot of devices have been developed to get the rope into the tree, especially since tree climbing has become a sport. The rope can be installed with a pruner or roping tool. A common option is to install a small rope that will be used later to pull up the climbing rope. Small ropes can be installed with a throwing ball, a throw bag, a special line gun, a special slingshot, a bow and arrow, or a small coil (monkey fist) of rope made by coiling the end of a rope into a bundle.
		The branches on some conifers are too dense to crotch a rope from the ground. A climber may have to carry the climbing rope up to the point where it is to be crotched.

(continued)

16.4.1 Tree Climbing Procedure *(continued)*

Step	Action	Details
5	Prepare fall protection.	Belaying is a fall protection system adapted from rock climbing that is very applicable to tree climbing. The system shown in Figure 16–2 is a belaying technique; virtually no extra equipment is needed because the climbing rope is used for fall protection and, later, for work positioning. This method can be used anytime a tree must be climbed, no matter which climbing method is used.

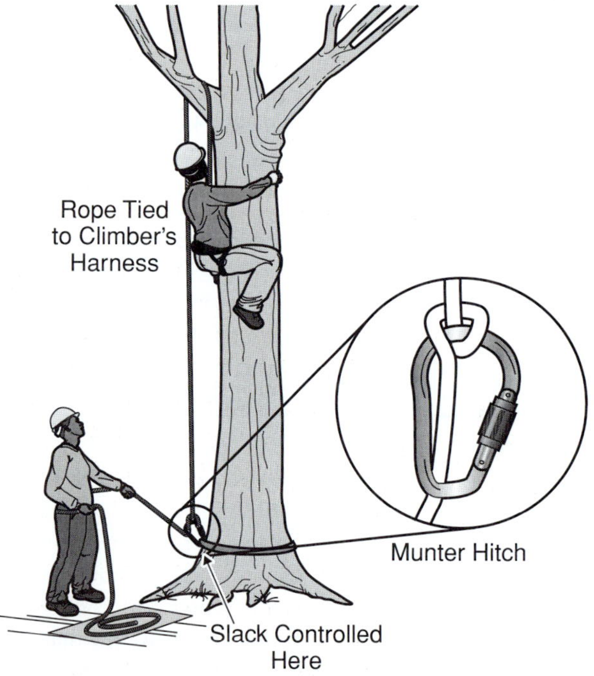

Figure 16–2 Tree-climbing fall protection.

After the climbing rope is in the tree, the climber attaches one end of the rope into both D-rings of the saddle and locks the keeper. A carabiner is attached to an anchor point (a tree) with a sling or is attached to both D-rings of a second arborist acting as an anchor. The climber should not be more than 25 lbs. (11 kg) heavier than the anchoring arborist.

(continued)

16.4.1 Tree Climbing Procedure *(continued)*

Step	Action	Details
		The free-hanging part of the climbing rope is hitched to a carabiner with a munter hitch (see Figure 16–3), and the carabiner keeper is locked.

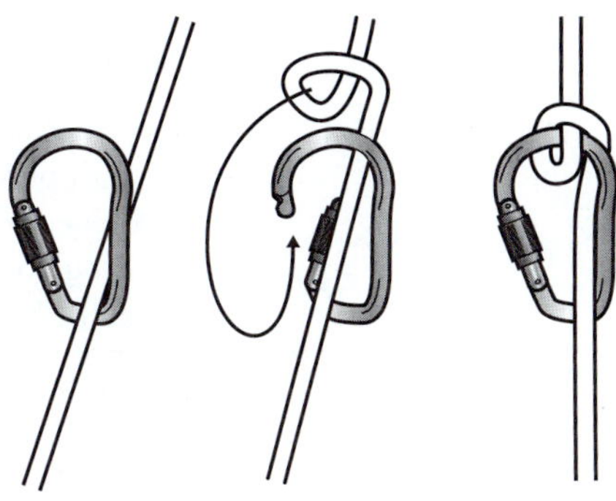

Figure 16–3 A munter hitch.

Step	Action	Details
6	Climb to first crotch.	Climb to the first crotch using any of the various climbing methods available to reach the lower branches. As the climber ascends, regardless of the climbing method, the anchoring arborist pulls the slack through the anchoring carabiner, thereby protecting the climber from falling to the ground.
		The length of the trunk and the equipment available will determine which climbing method would be most suitable. Sectional ladders are often used to get within reach of the lower branches to climb a tree.
		Tree spurs are not used very often anymore because the gaff wounds create entry points for insects and diseases. They are still used on trees that are to be cut down or on some trees that have an outer bark thick enough to withstand the gaff wounds. The gaffs on tree spurs are longer than for pole climbing, but there is always some doubt when climbing on heavy bark about the bark giving way.
		Shinning a short distance to get to the lower branches is not an uncommon practice but should be reserved for small-diameter trees and limited to less than 15 ft. (4 m). Climbers should be tied in to a belay system anytime they are climbing, but it is

(continued)

16.4.1 Tree Climbing Procedure *(continued)*

Step	Action	Details
		especially necessary when climbing with spurs or shinning.
		Body thrusting up a tree using a friction hitch on the climbing rope to maintain each gain in height is a skill that must be learned and practiced. This could be called a "self-belay method."
		Some mechanical ascenders are available that are rope-gripping devices that allow a climber to move up a rope. When using limbs to climb, hands and feet should be on different limbs and dead limbs should be broken or cut off on the way up.
7	Reposition if necessary.	When the first crotch is reached and it is necessary to go higher, the climber puts the lanyard around a main stem and sits back in the arborist saddle. Two hands are available to raise the rope to the next (or final) crotch.
8	Tie a friction hitch.	When the highest crotch is reached, the climber puts the lanyard around a main stem, and then the rope for belaying can be prepared for use as the climbing rope, with a friction hitch attached to allow work while suspended by the rope and seated in the arborist saddle. Depending on the climbing method, the arborist will tie the saddle onto the climbing rope with a friction hitch before leaving the ground or after reaching the intended top position in the tree. Historically, the *taut line hitch* (Figure 16–4) has been the hitch that holds an arborist in a working position and allows a controlled descent to other working positions and to the ground. It is a rolling hitch, so it is important to tie a half hitch or a stopper knot in the tail so it will not roll out and come undone. It is easy to slide it down with minimum pressure with one hand to control the speed of descent.

(continued)

16.4.1 Tree Climbing Procedure *(continued)*

Step	Action	Details

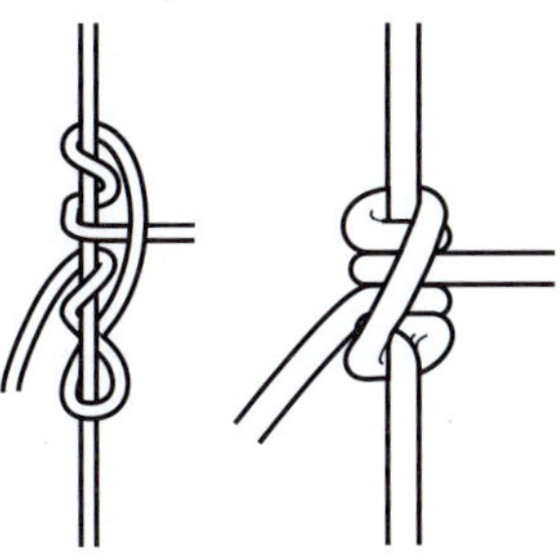

Figure 16–4 A taut-line hitch.

Two other friction hitches that can be used are the prusik and the Blakes. The prusik hitch (Figure 16–5) will not roll out but must be kept tight so it will grab. It can get too tight and will need two hands to get it to slide. The Blakes hitch (Figure 16–6) is the newest hitch and is becoming more popular. It is a little more complicated to tie and must be tied properly to hold. It can be tied with the free end of a rope and is self-adjusting. Where the tail passes through, the bottom two turns can become a hot spot that will melt if an arborist descends too quickly.

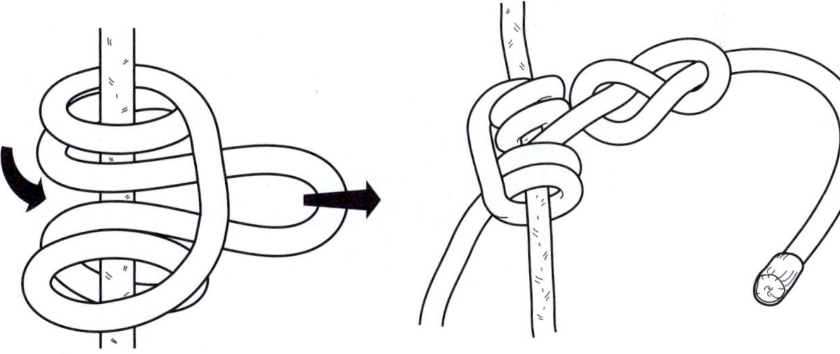

Figure 16–5 A prusik hitch.

Figure 16–6 A Blakes hitch.

(continued)

16.4.1 Tree Climbing Procedure *(continued)*

Step	Action	Details
		If transferring from an aerial bucket to a tree, an arborist should tie into a climbing rope with a friction hitch, anchored to a crotch in the tree before removing the lanyard attachment to the boom.
9	Work and descend.	The arborist can descend and stop along the length of the rope and carry out work. The climbing rope, friction hitch, and saddle comprise a work-positioning and fall protection system. If it is necessary to re-crotch while in the tree, the arborist uses a lanyard to tie in to the tree while tying into a new crotch.

16.4.2 Treetop Rescue Procedure

In most jurisdictions, occupational safety regulations require that when working in the vicinity of live circuits over 750 volts, a second person must be present within voice or visual range and trained to perform a rescue. The rescuer must also be trained in first aid and cardiopulmonary resuscitation (CPR). A rescuer must know and have practiced how to lower a bucket using the lower controls and how to get a victim out of a bucket.

Step	Action	Details
1	Call loudly to the injured employee.	At the first indication of a problem, call loudly to the worker aloft "Are you okay?" If the casualty responds but seems stunned or dazed, a rescue is probably needed, but the timing will not be as critical because the victim is breathing. Conscious victims of an electrical contact will often say they can descend on their own. Try to convince such victims to wait for a rescuer to climb up and act as a backup so that the descent will be controlled. *If there is no response, timing is critical.*
2	Call for help.	At the first indication that a rescue must be performed, call for help using a prearranged call or code words on the company radio or get help from

(continued)

Step	Action	Details
		nearby observers. The utility/contractor should have prearranged code words in place that immediately take priority over all other radio traffic.
3	Evaluate the situation.	If the casualty is still in contact, call for system control or dispatch to drop the circuit, preferably using prearranged code words to avoid a lot of discussion. If that does not work, the situation is such that the casualty is energized and the tree also may be energized. Climbing the tree presents a risk to the rescuer. Difficult as it may be to accept, from the ground it may be obvious that a rescue is not possible. If the casualty is in a precarious position, within the minimum approach limit, decisions regarding the need for rubber gloves, an insulated pruner or pole saw, or clean, dry rope will be needed to clear the victim. If the casualty is clearly not in contact, proceed with the rescue.
4	Climb up and attach a lanyard near the victim.	The rescuer climbs the tree with climbing equipment. Fall protection for the rescuer will depend on the tree, the practicality of getting a rope over a crotch quickly, the suitability of using the fall line of the casualty's climbing rope, the availability of work ropes, and the acceptability of a free climb risk. Climb to a position near the casualty and tie in with the lanyard. If the casualty is still in contact or too close to the line to approach safely, use an insulated pruner or pole saw to push or pull the casualty away from the powerline into a position that will allow the rescuer to approach the casualty.
5	Assess the casualty's condition.	If the casualty is breathing and/or conscious, the speed of the rescue is less urgent. Reassure the casualty while preparing to lower him or her to the ground. If a casualty insists on descending solo, the rescuer should insist on descending the tree with him or her. If the casualty is not breathing, it is urgent to get oxygenated blood to the brain as soon as possible. The best place to do that is on the ground. Before lowering

(continued)

Step	Action	Details
		a victim, some utilities/employers specify four quick mouth-to-mouth breaths to fill the lungs, some specify only lowering, and others specify starting CPR aloft.
6	Rig for lowering the casualty.	Different methods are used. It is important to use and practice the method specified by your utility/employer. One method is for the rescuer to crotch in a position above the casualty. The rescuer then descends to the casualty and ties the casualty's saddle to his or her own saddles by either hooking together their carabiners, tying a rope through the two trees of both saddles and connecting them together (using the end of the casualty's fall line) as shown in Figure 16–7.

Figure 16–7 Tree rescue.

Step	Action	Details
7	Lower the casualty to the ground.	Using the rescuer's friction hitch, the rescuer and the casualty come down together. If more trained workers are present, a worker on the ground could rig up and belay the two people to the ground.
8	Apply first aid and/or CPR.	When the casualty reaches the ground, apply first aid and, if needed, CPR (Look, Listen and Feel, 2 breaths, 15 compressions, check pulse).

16.5 Felling and Pruning

16.5.1 A Typical Tree Felling Procedure

Step	Action	Details
1	Prepare and discuss a safety job plan.	The plan should identify high-risk hazards and barriers needed to protect from the hazards. High-risk hazards include nearby powerlines, climbing hazards, hangers, dead wood, a restricted escape route, and a restricted drop area for a tree. If in doubt about the tree condition, use an increment bit to help diagnose the tree.
2	Make preparations to control the direction of fall.	Acceptable use in a forest may include notching in the desired direction of the fall, varying the thickness of the hinge on one side to control the fall, and/or using a wedge in the backcut, but near powerlines and buildings a tree should be winched, roped to a vehicle capable of pulling, or supported by a crane. Any tree that cannot be felled directly away from a powerline should be taken down piece by piece. Starting at the bottom, limbs are removed and work proceeds toward the crown, as in pruning. If the trunk cannot be felled, it is also removed in pieces. Work on problem trees may require skilled arborists and specialized equipment to carry out tasks such as climbing, working aloft, roping, piece-by-piece removal, or using a crane or winch. Problem trees are those that lean the wrong way, are hung up, have heavy tops, have broken or dead limbs (a "widowmaker" or a "chicot"), have a major fork in the trunk, have a falling path that is hard to predict (a "schoolmarm"), and/or have a restricted drop path. Be sure that clearance in the intended direction is adequate for the tree to fall completely to the ground. A lodged tree is very dangerous.
3	Rope the tree.	When removing a tree near a powerline, always rope it.
4	Prepare a retreat path.	Common causes of accidents include being struck by the butt, rebounding limbs, and broken tops, so it is important to clear a safe work area around the base

(continued)

16.5.1 A Typical Tree Felling Procedure *(continued)*

Step	Action	Details
		of the tree. Remove limbs, underbrush, and other obstructions. The retreat path should be at approximately a 45-degree angle to the rear of the notch in case the tree kicks backward off the stump. If the tree is on a slope, the retreat path should be uphill.
5	Establish an "exclusion" zone.	An exclusion or danger zone is an area where everyone other than the person designated to carry out a high-risk task is kept at a safe distance. An exclusion zone should be established for everyone other than the arborist making the backcut. Because of the hazard of flying missiles, such as broken branches, etc., a typical exclusion zone should be twice the length of the tree height.
		If rigging is used, include the bight of a winch and the area near a snatch block as the exclusion zone.
6	Cut notch.	A notch is not normally needed for trees less than 5 to 8 in. (12–20 cm) in diameter. When needed, a notch is cut on the fall side of the trunk to a depth of one-quarter to one-third of the tree diameter. The face opening of a proper notch should be equal to the depth or about 45 degrees. Make the lower cut of the notch first to prevent the loose wedge of wood from pinching or bending the chain.
		An improper notch is one where the notch closes and the tree stops falling:
		1. When the notch closes, it causes the tree to be under strain. Because of the strain, the fibers separate and the tree begins to split. The tree continues to split until it breaks off, leaving a "barber chair."
		2. When the tree stops, the hinge or holding wood must be cut off to drop the tree. Unless the tree is roped, there will be little control of the direction of fall.
7	Make a backcut.	Check again to ensure that everyone other than the saw operator is outside of the exclusion zone. Yelling "Timber!" is a reminder to an experienced crew.

(continued)

16.5.1 A Typical Tree Felling Procedure *(continued)*

Step	Action	Details
		Make a felling cut or backcut on the opposite side of the notch, 2 in. (5 cm) above and parallel to the horizontal cut of the notch. Leave a 1-in. (2.5-cm) uncut portion (holding wood). The holding wood serves as a hinge and will help cause the tree to fall 90 degrees to the orientation of the holding wood.
		Results of improper backcutting:
		1. A backcut that is at the same level as the notch increases the risk of the tree butt kicking back off the stump. The saw operator should work on one side in case the tree butt bounces straight back.
		2. A backcut that is made too high may prevent the tree from falling and will have to be pulled down.
		3. A backcut that is below the notch may cause the tree to drop back on the stump, and the tree may have to be pulled off the stump.
		4. A backcut that is too deep will cause the hinge to break and will result in little control over the direction of fall.
		5. A tree leaning heavily into the direction of the notch may create a *barber chair* as the backcut is being made. Cutting off the corners on each side of the notch before the backcut is made may help to prevent this from happening.
		Note the value of having a tree roped ahead of time. If a tree must be pulled over, the saw operator must be out of the exclusion zone.
8	Make your escape.	Saw operators escape or retreat while not turning away and losing sight of the falling tree because misjudged circumstances (wind, heavy crown) can cause a tree to fall in a different direction than planned.
		Give the tree time to settle into its final postion before approaching it.

16.5.2 A Summary of Tree Felling Hazards

Hazards	Controls
Falling dead wood can strike a worker when an overly mature tree or a mature tree such as oak, beech, or basswood is being felled. A dead top or limb that has fallen and is hung up, often between two trees, can fall on workers on the ground.	Too often, falling wood is an unidentified hazard and is called a widowmaker (chicot) for good reason. Inspect the tree for dead limbs and remove them first. If rotton wood is evident, inspect the whole tree. Use an increment borer or brace and bit to check for rotten wood.
Control of the falling direction of a tree can be lost. If the tree does not fall in the intended felling area, electrical contact is a possibility. Trees are electrical conductors.	Install ropes to ensure that a tree will fall in the intended drop area. The anchors for guide ropes must be secure. If a rope is used to pull the tree in a given direction, the vehicle pulling the tree must keep pace with the falling tree.

Hazards	Barriers
	If a tree becomes lodged on a live circuit, stay clear of the tree. If the tree cannot be pulled clear by the ropes already attached, or by a rope installed with a hot stick and rubber gloves, call for a clearance on the line.
There is no safe path for the tree to be dropped, or the planned felling area is too short for the height of the tree.	The radius of the felling area must be at least equal to the height of the tree. Everyone other than the saw operator must be outside of the "exclusion/danger zone" radius. Remove the tree piece by piece down to a level that allows it to be dropped safely. If climbing is necessary, use a secondary fall protection such as a belaying technique that requires training in the use of a carabiner and a munter hitch (Figure 16–3).

(continued)

16.5.2 A Summary of Tree Felling Hazards *(continued)*

Hazards	Barriers
A tree hung up on conductors or in other trees is under pressure and can fall or bounce in unpredictable ways or spring back when it is cut free.	Do not work in the presence of a hung-up tree. Have hung-up trees pulled down by a rope, winch, and vehicle.
Dead stumps, the wedge cut out from the notch, or other objects in the felling area can become dangerous flying missiles when a felled tree drops on them.	Remove loose objects from the intended drop area.
The saw operator is in danger if the tree butt kicks backward.	A clear escape route must be present at an angle of approximately 45 degrees to the rear of the notch. Never escape directly away from the direction of the fall in case the tree kicks backward off the stump. On a slope, escape in an uphill direction.

16.5.3 Pruning Guidelines

- Directional pruning (also called natural target pruning, drop crotch pruning, or the Shigo method) has become fairly standard for pruning near powerlines. Only the branches that head toward the powerline are pruned; those growing down or away from the line are left to grow.

- A hanger on a conductor can be safely removed from the conductor with an insulated hydraulic or manual pruner or saw.

- A saw cut should use the 1-2-3 method of pruning. A pruner cut on limbs that could split or cause the bark to strip should be pruned using the 2-3 method, which is the same as the 1-2-3 method without the undercut. Cut 2 will remove the weight, and cut 3 allows a clean cut in the desired location.

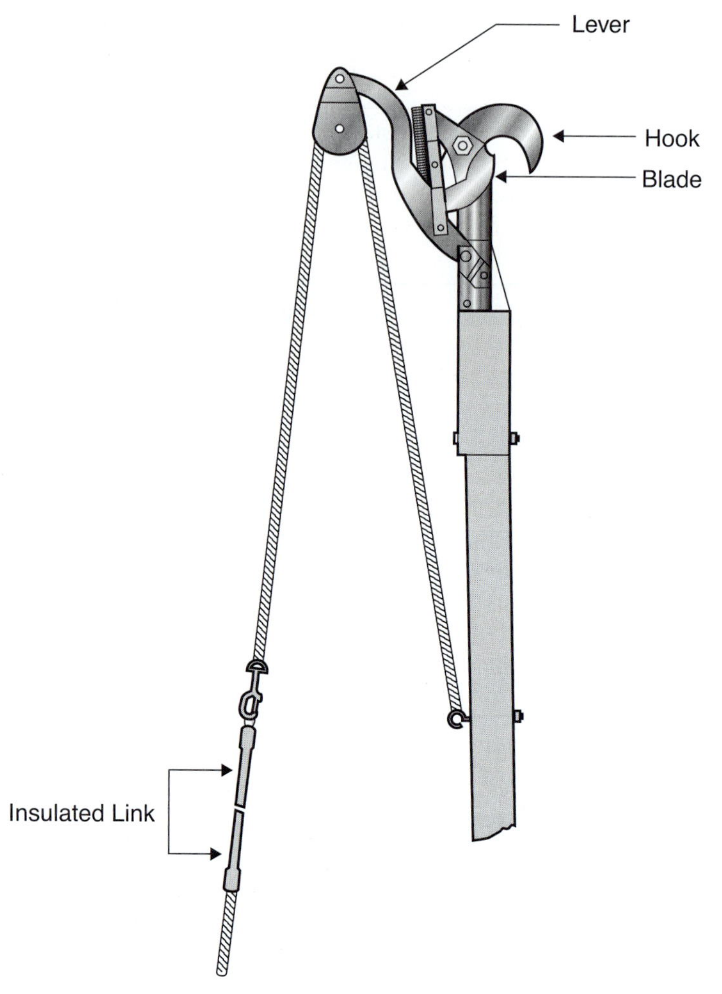

Figure 16–8   A pruner with insulated link.

- Avoid topping and leaving stubs. If a branch must be shortened, it should be cut back to a lateral that is large enough so that at least one-third of the branch diameter is removed, as shown in Figure 16–10.

- Many cuts may be needed to shorten branches hanging over power conductors. Each cut should leave the branch short enough that it cannot bridge phase to phase when dropped.

- When a larger limb is to be removed, the limb should be roped or rigged with rope blocks.

- At the end of any pruning job, every tree should be inspected for hangers that may become hazards to the public.

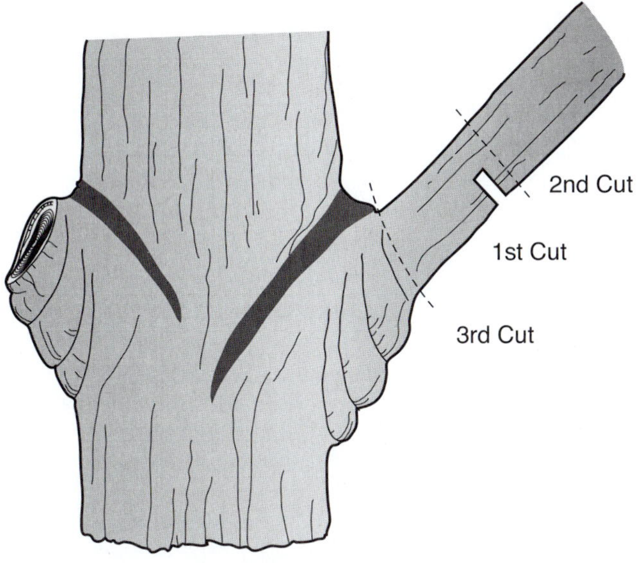

Figure 16–9   The 1-2-3 method of branch removal.

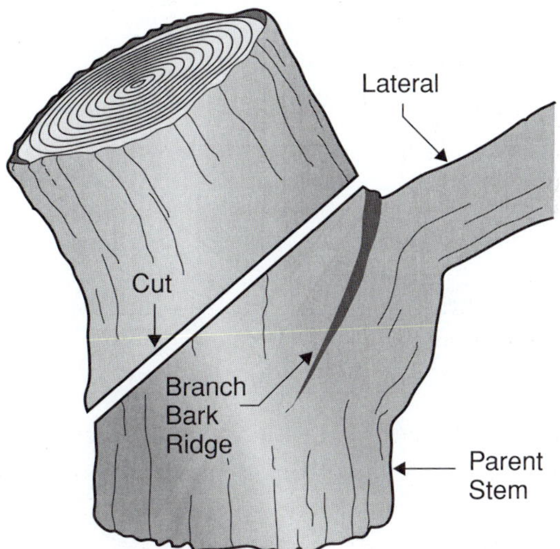

Figure 16–10   Cutting a leader.

CHAPTER 17

Entering, Inspecting, or Operating in a Substation

17.1 Entering a Substation

Most substations are remotely controlled. System control typically requires notification before anyone enters. If alarms go off in a control room, system control would like to know if people are there. Corporate safety rules and labor laws allow entry to control rooms only to people who have had electrical awareness training or are under escort by a qualified worker. One reason a substation has a fence, barbed wire, and warning signs is because, unlike powerlines, the clearance from live parts to ground or to a structure is easily reached with a boom, a ground rod, a ladder, and so on.

17.2 Hazards Working in a Substation

Hazard	Barrier
Contact with energized overhead bus or leads going into equipment bushings and potheads	Live apparatus is much closer than apparatus encountered in line work. Minimum approach distances for truck booms and workers are easier to bridge. Long objects, such as ladders, conduits, ground rods, and reinforcing steel, should be carried by two people, horizontally, when in the live part of the station.
Identifying the wrong equipment as having been deenergized	Substations contain a lot of duplicate equipment. Accidents have occurred because people have climbed on the wrong breaker. Positive identification with grounds and warning tape and signs is required.
Contact with induced voltage in live substations and close to overhead conductors	Any components sitting on insulators must be treated as hot until grounds are installed.
Contact with live underground cables when digging in a station	In existing stations, all powerline cables must be located and nearby digging must be done by hand or a "sucker" truck.
Opening a grounding connection can interrupt a current flow, and it can create a high voltage across the open point.	Ground connections should not be removed while a substation is energized. If a connection must be replaced, a temporary jumper should be installed.
SF_6 is heavier than air, so it will displace oxygen and create an immediate suffocation hazard, as would other heavier-than-air gases.	SF_6 gas is odorless, tasteless, and nontoxic in its normal state, but an SF_6 gas leak in a basement will drift down to the lowest part of the building and displace air. To reduce the risk of a large volume of SF_6 escaping ensure that SF_6 cylinders are stored outdoors and chained so that they cannot be damaged in a fall.
Arcing inside switchgears will decompose SF_6 gas. Decomposition products are considered toxic.	The decomposed products are metal fluorides and look like white powder. They can cause a rash, skin irritation, and eye irritation. Inhaling can damage the respiratory system.

(continued)

17.2 Hazards Working in a Substation *(continued)*

Hazard	Barrier
	Protective clothing and a full-face supplied-air respirator should be worn when entering electrical equipment. Powdered arc products should be removed with proper cleaning solvents and a vacuum cleaner equipped with a special in-line filter. Smoking, welding, or switching will cause decomposition of any escaped SF_6 into toxic material.
Damage to a control cable may prevent a critical piece of equipment from operating.	There may not be an immediate reaction to damage to a control cable. However, the damage will get worse over time and the control cable will eventually fail. This can result in equipment failing to operate at a critical time because it will not get the signal to operate. Call for repair if damaging contact is made with a cable.

17.3 Switching in a Substation

Many substations are remotely controlled, but individual components in most stations must be operated by a person.

A powerline worker may be asked to operate switchgear in a substation as an agent under the direction of system control. Using an operating drawing or a single-line diagram to follow as instructions are given by system control is essential. Figure 17–1 is a sample single line drawing.

When a switch, with a handle, is operated by someone standing on the ground, an insulator failure can cause the frame of the switch and structure to become energized. In a substation, each of these types of switches should have a metal grid or platform that is electrically bonded to a switch handle. The grid may be buried under gravel. This equipotential arrangement ensures that an operator's feet and hands will be at the same potential when things go wrong. Rubber gloves alone are not adequate protection.

If asked to open a switch with power fuses, be aware that these devices are not designed to drop load. If asked to open or close a recloser or breaker from a control box mounted right at the equipment, be aware that a breakdown within the recloser or breaker can cause a fault to ground and energize the grounded network. Use a portable ground-gradient mat if a permanent one is not installed at the control box.

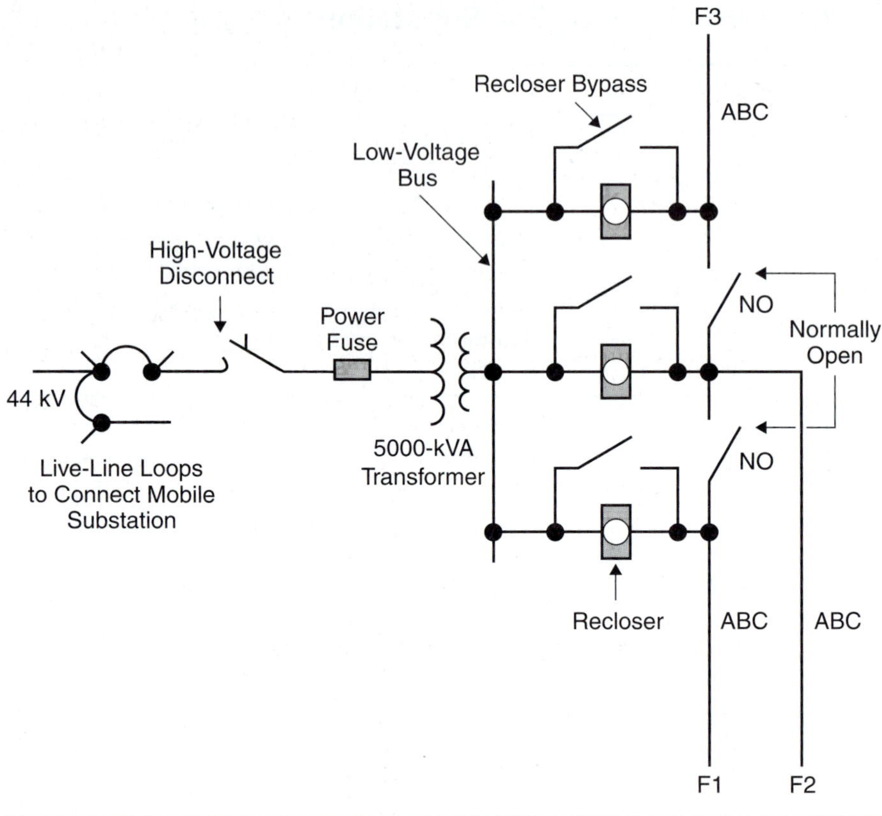

Figure 17–1   A single-line diagram for a simple rural substation.

17.4 Inspections

Inspections and records for the condition of all substations are required by legislation in most jurisdictions.

Types of inspections

1. Monthly visual inspection

2. Annual inspection with diagnostic online test equipment

3. Annual detailed inspection of major equipment

4. Internal inspection of transformers and regulators, when required

17.5 A Walk-Through Substation Inspection

The following walk-through inspection log is a guide for people who are not substation maintenance workers. Employers typically have their own forms, probably a digital instrument where the input is logged into the utility work management system.

Item	Inspection	Findings
The yard	Check: • For holes and gaps under the fence. A gap should be less than 2 inches under the fence and 4 inches under the gate. • That the barbed wire is taut. • That there is no third-party metallic fencing tied directly to the substation fencing. • Locks and security of gates. • That no material is stored within 5 ft. of the inside or outside of the fence. • That the braid-type ground connections are secure. • That all grounding mats are covered.	
The transformer	Depending on transformer size, available gauges, and so on, look for: • Damaged bushings. • Oil leaks and the oil level in the main tank and tap-changer compartment. • Both the oil and winding temperatures reading below 70°C (under normal operating conditions). The maximum winding temperature is stated on the nameplate. • Leakage, proper pressure, etc., of the inert gas system (when applicable). • The operations counter indicator reading associated with the load tap changer, and read it.	

(continued)

Item	Inspection	Findings
Tap changers and voltage regulators	The operations on the load-tap changer are read and recorded from the counter indicator. Inspection of a voltage regulator is similar to that of a transformer as far as oil leaks and so on go. More detailed inspection might have the regulator control put in the manual position and operated up or down over a small range and then returned to automatic to watch it go back to the former position. Each of the three phases of the voltage-regulator taps should be within four positions of each other.	
Oil, vacuum, SF_6, and air-blast circuit breakers	Check for loose, contaminated, or damaged bushings; loose terminals; oil leaks; and proper gas pressures. Check oil level in bushings and main tank (as applicable). Check anti-condensation heaters. Read and record the number of operations on the indicator.	
Capacitors	Look for damaged tanks and bushings, and for any leakage of dielectric fluid.	
Reactors (oil-filled and air core)	Inspection is similar to the transformer inspection.	
Disconnect switches	Check for cracked, contaminated, or broken porcelain; loose connections; and corrosion to metal parts.	
Reclosers	Inspect for oil leaks. Recloser settings can be checked, where applicable, to ensure that the ground trip, reclosing, and supervisory settings are in the desired positions.	
Surge arresters	• Check for damaged porcelain. • Check for loose connections. • Check for pitted or blackened exhaust parts.	

(continued)

Item	Inspection	Findings
	• Record the number of operations, if the units have discharge counters.	
Batteries	Check that the batteries are at about 125 volts, check that the charger is working, and check for leaks.	
Porcelain/ insulators	Check for chipped or cracked porcelain. Damaged glazing on a portion of the porcelain indicates flashovers.	
Capacitors	Check capacitor tanks for leakage or bulging.	
PTs and CTs	Check current and potential transformers for damage to cases, bushings, terminals, and fuses.	
Ground grid connections	Check visible ground connections at equipment, structures, switching platforms, and fences. Check continuity of fence grounds.	
Meter readings	Take any requested meter readings.	

17.6 Typical Tests Carried Out in Substations

When a visual check is not enough, numerous tests can be done. The test descriptions below are only intended to alert a powerline worker to the tests that are available and do not take the place of the training and procedures needed to carry them out safely.

Test	Description
Infrared Thermography	Infrared thermography can be used to survey a substation for hot spots. Temperature differences indicate needed actions: • 2°C to 4°C indicates a need to investigate. • 4°C to 15°C indicates a need to schedule a repair. • 16°C and above indicates a need for immediate repair.

(continued)

Test	Description
Insulation Resistance Test	Using a 500/1,000/2.500-volt megger, this test is performed to verify the integrity of the insulation and can be used on transformers, circuit breakers, cables, motors, switches, etc. *Some typical minimum resistance values* Motors, 75°C 120-volt–1.12 megaohms 208-volt–1.20 megaohms 240-volt–1.25 megaohms 480-volt–1.50 megaohms DC motors, 75°C all voltages–1.00 megaohms High-voltage circuit breakers, 20°C 10,000 megaohms High-voltage bushings, 20°C 10,000 megaohms
Insulating Oil Tests	Oil samples are typically sent to a laboratory where they are checked for neutralization number, interfacial tension, dielectric strength, gas analysis, and color test. Results are compared with previous tests to see if any trends are developing. Reclaim and/or recondition the oil as follows: • *Recondition* when neutralization (acid) number (acids are formed as oils oxidize and age) exceeds 0.35 to 0.5 for transformers or for circuit breakers. • *Reclaim* when interfacial tension (measures the surface tension of the oil against that of water) is less than 18 dynes/cm. • *Recondition* when the dielectric strength is less than 28 kV with a standard gap. • *Recondition* when the color exceeds 3.5. • *Recondition* when the oil is cloudy, dirty, or contains visible water.
Contact Resistance Test	A low-resistance ohmmeter can be used to measure the resistance of the main contacts of a circuit breaker and determine if there is a need for maintenance of the contacts. *Typical resistance readings,* Reclosers: 50–200 micro-ohms Medium-voltage circuit breaker: 10–50 micro-ohms High-voltage circuit breaker: 50–350 micro-ohms

(continued)

Test	Description
DC High Potential Test	This test is similar to that used on underground cables as a test of the dielectric strength of insulation. It is most suitable to test solid insulation, such as, porcelain, rubber, PVC, and PE. The test voltage has to be chosen carefully so that the insulation is not damaged.
Power Factor Test	The power factor test measures the ratio between capacitance and resistance within insulating material. It is most suited to finding voids. Test values are compared with values read at different times and recorded to establish trends with previous readings. Acceptable values of power factors for various pieces of equipment may come with the instructions with the test equipment and/or are available from equipment manufacturers.
Dissolved Gas Analysis	Combustible and noncombustible gases are produced over time by arcing faults or small turn-to-turn arcs. Fluid samples are generally sent to a laboratory for gas analysis. Tests that indicate 0.5% to 5.0%, by volume, of combustible gases indicate the probable beginning of a fault.
SF_6 Gas Analysis	During the interruption of current, a circuit breaker will contaminate the SF_6 gas insulation medium in SF_6 circuit breakers. Instead of testing, SF_6 gas should be reconditioned on a regular basis.
Dielectric Absorption Test	Using a 2,500-volt motor-driven megger or similar device, the quality of insulation of transformers, regulators, and shielded high-voltage power cable is tested by leaving the equipment/cables fully charged over time. Typically, resistance readings are taken every 15 seconds during the first 3 minutes and then at 1-minute intervals. This continues until three equal readings are read and recorded to establish trends with previous readings.

Test	Description
Timing Test and Motion Analyzer Test	The timing test verifies that all poles of the circuit breaker and all series contacts in each pole operate simultaneously. The motion analyzer test verifies the condition and proper adjustment of the mechanical operating linkages of a circuit breaker.
Winding Resistance Test	The resistance of a transformer, regulator, or motor winding should be the same as it was during the original acceptance test. Any change, excluding normal change because of the temperature during the test, is an indicator of the beginning of trouble.
Turns Ratio Test	Periodically perform a turns ratio test as an aid in detecting turn-to-turn short circuits in power and instrument transformers.
AC Over-Potential Test	This test applies about 1.7 times the operating voltage on devices and associated wire rated at 600 volts or less. The test verifies the integrity of the insulation. It is best to really know the limitations of this test before applying it to solid-state components.
Over-Current Tests	Reclosers and low-voltage circuit breakers can be tested to verify the minimum instantaneous trip current level, as well as the calibration and proper operation of the device.
Ground Grid Resistance Measurement	The ground grid resistance is measured periodically to help detect any changes from previous measurements. Any significant changes are investigated.

APPENDIX A

TABLE A–1   Calculations Involving Power

To Find	Direct Current	Single-Phase AC	Three-Phase Wye AC
Kilowatts	$\dfrac{I \times E}{1000}$	$\dfrac{I \times E \times pf}{1000}$	$\dfrac{I \times E \times 1.73 \times pf}{1000}$
Kilovolt-Amperes (kVA)		$\dfrac{I \times E}{1000}$	$\dfrac{I \times E \times 1.73}{1000}$
Amperes (when kW are known)	$kW \times \dfrac{1000}{E}$	$kW \times \dfrac{1000}{E \times pf}$	$\dfrac{kW \times 1000}{1.73 \times E \times pf}$
Amperes (when kVA is known)		$\dfrac{kVA \times 1000}{E}$	$\dfrac{kVA \times 1000}{1.73 \times E}$

The preceding table gives formulas used to calculate power in three-phase and single-phase circuits. Calculations for field applications use the kilovat-ampere formulas.

Calculations involving three-phase power use line-to-line (phase-to-phase) voltage.

The square root of 3 is normally used at the value of 1.73.

A.1.1 Calculations with "Handy Numbers"

To be able to do the preceding calculations quickly in the field a "Handy Number" can be used for approximations.

Approximate kVA = Amperes × "Handy Number"
Approximate amperes per phase = kVA ÷ "Handy Number"

TABLE A–2   Examples of "Handy Numbers" for Some Voltage Systems

Calculating Approximate Loads with "Handy Numbers"

	3-Phase kV	"Handy Number"	1-Phase kV	"Handy Number"
kVA = Amps × "Handy Number" Or Amps per Phase = kVA/"Handy Number"	230	400		
	115	200		
	69	120		
	46	80		
kVA = Amps × "Handy Number" Or Amps per Phase = kVA/"Handy Number"	25	40	14.4	14
	12.5	22	7.2	7
	8.32	14	4.8	5
	0.208	0.36	0.12	0.12

To calculate the "Handy Number" for other voltage systems:

- For three-phase lines, the "Handy Number" = ∅-to-∅ voltage × 1.73 ÷ 1000
- For single phase, the "Handy Number" = ∅-to-neutral voltage ÷ 1000

A.1.2 Examples Using "Handy Numbers"

Question: One phase of an 8.3/4.8 kV feeder has 80 amperes more load than the other two phases. To balance the feeder, how many kVA should be transferred to the other two phases?

Answer: Using a "Handy Number," which is 5 for a 4.8-kV single phase line, × 80 amperes = 400 kVA. Therefore, 200 kVA should be transferred to each of the other two phases.

Question: Load on each of the three phases of a 120/208, 50 kVA pad-mount transformer bank are 150 A, 170 A and 140 A. Is the transformer overloaded?

Answer: The average load on the three phases is 150 + 170 + 140 ÷ 3 = 153 Amperes. Using the "Handy Number," which is 0.36 for 120/208 volt service × 153 = 55 kilovolt-amperes. The transformer is only slightly above it's rating, but the load should be balanced more between the three phases to prevent one transformer winding from being overloaded.

TABLE A–3   Metric Conversion: Linear

To Convert	Multiply By	To Obtain
inches	25.4	millimeters
inches	2.5	centimeters
feet	0.3	meters
yards	0.9	meters
miles	1.6	kilometers

TABLE A–4   Metric Conversion: Weight

To Convert	Multiply By	To Obtain
fluid ounces (US)	28.4	grams
pounds	0.45	kilograms
ton	0.9	tonne

TABLE A–5   Metric Conversion: Volume

To Convert	Multiply By	To Obtain
fluid ounces (US)	29.6	milliliters
1 gallon (US) (8.3 lb)	3.8	liters
1 gallon (imperial) (10 lb)	4.5	liters
1 quart (US)	0.95	liters
1 quart (imperial)	1.1	liters
cubic yard	0.8	cubic meters
cord (128 cu ft.)	3.6	cubic meters

TABLE A–6   Metric Conversions: Miscellaneous

To Convert	Multiply By	To Obtain
Btu/hr	0.2931	watts
Horsepower (hp)	746	watts
Pounds per sq. in.	6.9	kilopascal (kPa)
acre	0.4	hectare

TABLE A–7   Speeds

Speed of Light	Speed of Sound
300,000 km/sec. 186,000 mi/sec.	11,00 ft/sec. = 382.8 m/sec. = 856.3 mi/hr. = 1378.1 km/hr.

TABLE A–8   Temperature Conversions

From Fahrenheit to Celsius	From Celsius to Fahrenheit
To convert from degrees Fahrenheit to degrees Celsius, subtract 32 degrees from the temperature and multiply by 1.8.	To convert from degrees Celsius to degrees Fahrenheit, multiply the temperature by 1.8 and add 32 degrees.

TABLE A–9   Metric System Prefixes

Tera	1,000,000,000,000 (10^{12})	Deci	0.1 (10^{-1})
Giga	1,000,000,000 (10^{9})	Centi	0.01 (10^{-2})
Mega	1,000,000 (10^{6})	Milli	0.001 (10^{-3})
Kilo	1,000 (10^{3})	Micro	0.000 001 (10^{-6})
Hecto	100 (10^{2})	Nano	0.000 000 001 (10^{-9})
Deca	10 (10^{1})	Pico	0.000 000 000 001 (10^{-12})

TABLE A–10   Geometric Figures

Area of a Circle	Circumference of a Circle	Area of a Triangle	Area of a Sphere	Volume of Sphere
$\pi\, r^2$ ($\pi = 3.14$)	$\pi\, D$ or $2\,\pi\, r$	Base $\times\ \frac{1}{2}$ Height	$\pi\, D^2$ or $4\,\pi\, r^2$	$D^3 \times 0.5236$

Power Line Work Lingo

There are many terms for the work methods and tools being used for Power-line work. Some regional terms used by linemen on discussion groups are not understood by all of us.

Aerial Device—aerial bucket, bucket truck, aerial lift, cherry picker, aerial-lift platforms,

Alive—hot, energized, alive, heat it up

Anchor—dead man

Apprentice Lineman—mud flap, trainee, learner, grunt, helper, squeek, chump, dipstick, no class, whistle-pig

Auxiliary Arm—lay out arms, hot arm

Backfeed—{Live secondary, because of generator, jumpered bus break, extension cord from neighbor's house feeding back through transformer and energizing primary}.

Bolt Cutter—Pie Knife

Bracket Grounding—Box grounding,

Bull Wheel—butterfly

By-Pass—shoo-fly, a temporary line by-passing the main line to allow dead line work on the main line.

Cant Hook—gut wrench, hook, peavy

Capstan Hoist—cat head

Chain Hoist—guy jack, coffin hoist, strong arm

Chain Saw—sander, gas axe,

Clamping in—clipping in, tying in

Clearance (for work)—work permit, lockout/tagout protection, LOTO, outage, mark up

Confined Space—enclosed space, restricted space

Connector—bug, fargo, blue dot, double deuce, deuce bug, 336, big bug, green dot, single bolt, double bolt, #6 bug (all depending on the size)

Controlling Authority—dispatch, system control, controlling operators, control room

Daylighting—potholing

Deenergized—dead, cold wires

Digger/Derrick—line truck digger, radial boom derrick (RBD), standard line truck digger (SLTD), electric line truck.

Do Not Operate Tag/Card—Do Not Connect Tag, Do Not Energize Tag, Keep Your Hands Off Tag

Hold Tag, Locked Out Tag, My Life Is On the Line Tag—Do Not Reclose Tag—Hold Off Tag, Stand Off Tag, Hold Tag, Hot-Line-Hold (HLH)

Drill—brace & bit, saw dust pump

Elbows (Load Break and Non Load Break)—pistol grips

Electrician (indoor)—narrowback

Energize—light, light up, liven up, power up, going hot, heat it up

ESR Footwear (Electrical resistant footwear)—DP (Dielectric Protective Footwear), Shock Resistant Footwear

Extension Link—dillibob (Uually an 18" fiberglass link, used when a single phase primary is coming off of a three phase pole w/crossarm, it extends the single phase line out past the three phase arm it goes between the eyebolt and the dead end insulator, epoxillator), donkey-dick

Flagger—flagman, flag person, traffic control person, road kill

Floating Dead-ends—openers, live line openers, slug, poor-boy

F.R. Clothing (Flame Resistant), Arc Resistant,

Fuse Chamber—door, cartridge, fuse tube, gate

Generator—leroy, perkins

Gin—pole buddy, old man, davit

Guy fitting—guy hook, goathead, bullhead, steerhead, guy attachment, p hook, pig ear, ram's horn, bear paw (not the donut)

Guy Strain Insulator—egg breaker, johnny ball, insul stick

Ground Gradient Mat—Ground mat, switch operating platform, ground plate

Groundman—helper, grunt

Ground Rod Clamp—peanut, acorn

Handline Hook—meat hook, pulley hook

Hazard—danger, widow maker,

Hog Wagon—a temporary transformer (enclosed in a small trailer with reusable leads)

Hot Pole—a pole where the insulators breaking down or cutout is leaking though the post to the bracket causing the pole to be energized.

Jaw Grip—wire grip, bulldog, cum-a-long, chicago grip, pork chop, widow, Dutchman jaw clamp, wire clamp,

Job Briefing—tailboard conference, tailboard briefing, toolbox talk, tailgate talk

Jumper—{One man jumper holder- parrot stand, solo holder} {Squeegee or a spring jumper, curly-q, its a piece of wire wrapped up like a coil with a hot line clamp at either end to make a quick jumper for bypassing a tap, also used on either a transformer or a cutout, although design engineers do not like it.}

Lag Screw—hardhead

Lateral Tap—spur, lateral, tap, primary tap, run off

LEL (Lower Explosive Limit)—LFL (Lower Flammable Limit)

Live Line Clamp—hot line clamp, tap clamp

Minimum Approach Limit—absolute limit of approach, {Breaking clearance" especially on transmission work, getting too close to danger}

Operating Drawing—one line drawing, single line drawing, operating map

Outriggers—stiff legs, stabilisers

Plastic Coverup—rigid coverup, tupperware, hard cover, barriers, taco

Pole—stick

Pole Down Ground—vertical,

Pole Platform—diving board, baker board, insulated pole platform, live line board, hot board

Pole Top Pin—lead head, bayonet, ridge pin

Pot Head—riser, terminators

Pole Shock—electric shock from a hot pole.

Potential Tester—jiggler, wiggle, pot tester, P.I.-Potential Indicator

Powerline Worker—lineman, lineperson, lineworker, line installer/repairer, linehand, powerline maintainer, line mechanic, powerlineman, power line technician, distribution line technician, line techs, powerline repairer, clum sum, electrical linepersons overhead lineworker, electrical lineworker, powerliner, GIB (guy in bucket), Scratch and Sniff crew (outage response crew) hiker; line servicer, bucket monkey, pole jock, pole cowboy, skywalker, {Supervisor's versions—coffee drinkin', knuckle draggin', over paid bucket monkey's}

Powerline Worker, Transmission—hi-liner, hi-line tech, static head,

Preform Grip—stringing sock, sugar grip, flexible mesh pulling grip, horse cock, sock line, kellum grip, chinese string, dog cock

Preform Thimble—horse collar, #1744, pig ear

Quick Sleeve—pickle

Pull Rope—stringing rope, bull line, pulling line, pulling rope, hard line

Primary Step-down Transformer—ratio transformer, rabbit

Reel Handler—Coca-cola (the reel thing), Finigan Pin

Rest Break—Skippy

Rigging Terms—tool blocked (can't take anymore on your rigging, have to rerig, should have thought of that before you rigged), blanket sling (tarp for setting poles)

Riser—stinger, lead, primary lead, sweep, siphon

Riser Pole—terminal pole, transition pole, dip pole, dig pole

Rope Ladder—jacob's ladder, flex ladder

Rubber Blankets—rubber covers

Rubber Dead End Cover—pig

Rubber Hood—insulator cover

Rubber Hose—line hose, hose, rubber, flexible protective equipment, eel, guts, reptile, snakes, rubber covers, coverup, slinky

Rubber Gloves—condoms

Rubber Sleeves—arm rubber

Secondary Cables—triplex, {cutting sections is called cribbing, splitting the crib, or pigtailing the ends, or cut back}

Secondary Fault—hotspot, holiday

Shield Wire—static wire, ground wire, counterpoise, sky wire

Shotgun—grip-all, clamp stick, grab-all, egg sucker

Sleeve—splice, (quick sleeve—pickle)

Solid Blade Switch—solid fuse, slug, bypass

Spiking Cable—spearing cable

Spacer, Aerial Cable—crow

Spun Bus—Wrapped Secundary, aerial cable

Spurs—hooks, climbers, westerns, lady slippers, sky hooks, hikers, gaffs, money makers, mortgage lifters, jumpers, jojos, leg irons,

Strain Insulator—dead-end insulator, suspension insulator, bell insulator, elephant ear, fish, four and a quarter, poly link,

Strand—messenger cable

Stringing Block—traveler, dolly, roller, stringing sheave.

Switch Stick—disconnect switch, hook stick, P2 Head, window pole

Switching—Arcing and Barking (closing in on fault)

Switching Order Form—Order-to-Operate, Operation Order,

Swivel Stick—chicken catcher (2 inch ring on end of switch stick used to install preformed tie on top tie or side tie)

Teasing—fuzzing, buzzing

Thumper—surge generator, impulse generator, capacitive discharge set, banger

Tag—card

Top Tie Insulator—bottle, chili bowl, post

Transformer—pot, can, trans, kettle, bug, stepper, tanks, tubs, tx

Vault—manhole, chamber, cable chamber, maintenance hole

Vibration Damper—dog bone, johnny ball,

Working Load Limit (WLL)—rated capacity, maximum working load, safe working load, safety ratio

Worksite Protection—traffic control

INDEX